Psychology and Societal Development

About the Author

Girishwar Misra, Ph.D., is Professor and former Head, Department of Psychology, and Dean Faculty of Arts, University of Delhi. He has been Fulbright Fellow at Swarthmore College and the University of Michigan. During his academic career spanning over three decades, he has previously taught at the Universities of Bhopal, Allahabad and Gorakhpur. The Government of Madhya Pradesh conferred upon him the Radhakrishnan and Dr. Hari Singh Gaur Awards. Professor Misra has authored or edited *Psychological Consequences of Prolonged Deprivation, Deprivation: Its Social Roots and Psychological Consequences, Applied Social Psychology in India, Perspectives on Indigenous Psychology, Psychological Perspectives on Stress and Health, Psychology of Poverty and Social Disadvantage*, and *Contributions to Psychology in India*. He has been President and Convener of the National Academy of Psychology, India. He is currently editing its journal *Psychological Studies. He can be contacted at* misragirishwar@gmail.com.

Psychology and Societal Development

Paradigmatic and Social Concerns

GIRISHWAR MISRA
Professor, Department of Psychology
University of Delhi
Delhi-110007

CONCEPT PUBLISHING COMPANY, NEW DELHI-110059

Cataloging in Publication Data—DK
Courtesy: D.K. Agencies (P) Ltd. <docinfo@dkagencies.com>

Misra, Girishwar, 1951-
Psychology and societal development : paradigmatic and social concerns/ Girishwar Misra.
p. cm.
Includes bibliographical references (p.)
Includes index.
ISBN 8180692892

1. Social psychology. 2. Social psychology—India. 3. Social change—Psychological aspects.
I. Title.

DDC 302.095 4 22

ISBN 81-8069-289-2

First Published 2006

Published and Printed by
Ashok Kumar Mittal
Concept Publishing Company,
A/15-16, Commercial Block, Mohan Garden,
New Delhi-110059 (INDIA)

Phones : 25351460, 25351794
Fax : 091-11-25357103
Email : publishing@conceptpub.com

Preface

This collection of essays on various aspects of psychology presents a reflective response to the issues concerning scientific paradigm, culture, and relevance encountered while engaging with psychology as a science and practice. The essays relate to the struggles, and voice certain concerns that emerge while dealing with the realm of human mind and behaviour. To put the essays in place I would like to share with the readers some of the background considerations that have gone into the development of ideas presented here.

The idea of psychology as a modern discipline has just completed 100 years. It was in 1905 that Dr. B.N. Seal, King George V Professor of Mental and Moral Philosophy at Calcutta University, drew up the syllabus of psychology after consulting the courses run at European and American Universities. It starts with the scientific spirit of the physical science tradition. The received scientific model views behaviours or other psychological process (e.g. ability, personality, interest, aptitude, attitude, need, value, motive, perception, learning, motivation) as observable and quantifiable entities like natural objects. It treats psychological processes as static and decontextualized and independent structures or entities with intrinsic meanings waiting to be discovered by the standardized procedures. This strategy has obvious advantages. It allows the interlocutors to wield authority and grants a sense of being scientific and therefore offering a privileged position in terms of the membership of a power group. The presumed objectivity and quantifiability allows certain degree of manoeuvrability that at the first glance is very impressive.

Upon serious reflection, however, it is realized that construing psychological reality within the tradition of natural sciences is just one of the views or vantage points but certainly not the only not the only view. It has certain inbuilt constraints and its metatheoretical frame of reference, portrayed as the right one, is increasingly coming under critical scrutiny. Theoretical advances particularly in physics, biology and the historiography of science and sociology of science have demystified the unquestionable objectivity, certainty, and rationality attributed to science, which happens to be just another human activity and subject to errors that are common to the human realm.

In the present context it is quite clear that the idea of observer independent reality is untenable and science is a social construction in which what counts as science depends on the community of scientists. Things that conform to the norms, rituals and practices accepted by the community of scientists at a given point of time and place define "science" and other things are excluded from it and treated as non knowledge. This kind of knowledge may be challenged but the challenge will not be accepted or entertained by those at the helm of affairs. Things may change if a change in the world view takes place or a scientific revolution occurs. The world view may change because of other happenings that are independent of the science in practice. In important respects science cannot be considered as an autonomous institution.

The natural science image that promises principles, free from the limitations of time and space, and its inheritors, find it odd to bring culture as an issue in the course of deliberations about psychological processes. They treat humans as a biological structure and feel that its functioning can and should be handled in terms of mechanical and reductionistic strategies because it is transparent, self-correcting and objective. It represents the received view of psychological science and for a large segment of the psychology community it serves as the undisputed truth. However, it misses the point that such a position undermines the complexity of being human in terms of its documented and verifiable range of experiences and phenomena. In this way it constricts the notion of human reality by imposing its own categories and declaring that as *the* reality. The method-reality dialectic is missed in the process and the robustness of sciences is used to legitimize the effort.

The practice of psychology as a natural science has paid its dividends. It has given a sense of scientism and apparent success in achieving the goals of understanding, explanation and prediction. It has also helped in attaining certain degree of control. Also, it has provided means to play the game of science. Many of us enjoy a sense of progress simply because of time. However, the model has not fared well in studying many problems and has raised additional problems. The myth of science as a new religion has been so overpowering that we have lost sight of these developments.

The picture of culture free or acultural scientific psychology is reinforced by adopting a disciplinary story that establishes Wundt and his tradition of experimental science as the representative model of the discipline and cultural concerns as secondary or a part of the problems to be explained. Historical investigations have revealed that such has not been the case. Not only did Wundt take culture seriously but human and interpretive sciences too have contributed to further our understanding in important ways.

Today we witness considerable diversity in the theoretical accounts of psychological phenomena that are rooted in different metaphors. The metaphors have changed with time. Thus the black box is now replaced by a computational metaphor. If we examine the developments in hard core areas like learning and memory we find that there is no cumulative progress in psychological knowledge. Instead we have raised new questions. Also, the image of human mind as brain activity presents only a partial picture of reality. The reductionistic exercise in psychology has not worked. Higher order qualities cannot be derived from lower order things.

The significance of culture in organizing human behaviour makes a strong case to evolve a culturally informed perspective on psychology. The psychosocial processes appear to be evolving and situated in cultural contexts. In case of human beings who use symbols and incorporate the outcomes of action the problem is more complex. The effect of knowledge on human beings makes this even more complicated. Once people know that they have certain characteristics, certain powers, certain attitudes, the mere knowledge of that may trigger change. The consequent change in a person's own understanding of who he or she is, happens to be a potential source of change in his or her psychology, naïve determinism simply does not work. The problem of subjectivity and personhood as part of one's experience also demands our attention. The male dominance of psychological theories also deserves analysis. The feminist critique of epistemology and methodology has emerged as a different voice that needs to be heard.

I often realize that the practice of the natural science model has been responsible for maintaining *status quo* in the discipline because the job of a scientist gets confined to the context of justification. He works more like a pathologist taking samples and determining their deviations from the standard. In contrast, the context of discovery is open to all. Anyone can contribute the hypothesis but its testing is the job of a scientist. Prototype of science or the scientific method focuses on a systematic appraisal of some idea or hypothesis. Its problems are experienced by many of us but are hardly attended to. The theories continue not because they are true or because they are not true but because we are affectively engaged with them and our social context demands them. The theories that stay and those that vanish from the scene, meet their respective fate on account of our affective investment.

It is quite clear that social sciences grow and are sustained by the matrix of social forces. They, therefore, need to relate to the social context and make disciplinary efforts socially relevant. The interface with social problems also raises conceptual issues that lead to theoretical innovations. Some of the essays included in this volume have tried to address related issues.

In any case the realities and our modes of work and engagement with them, however, are of our own making. While dealing with them we are quite passionate. We assign meaning to them, use them and denigrate or celebrate them. We continue with them or discard them. We use them and invest our ideas, resources and if required sacrifice for them. All this happens in a cultural context and culture figures centrally in our efforts.

The present volume elaborates the above ideas and the ways psychology has been changing its focus and how it has tried to respond to the demands of contemporary reality. Ideas and meanings are social products. Interactions, discussions, and interchanges on numerous formal and in formal occasions with my mentors, seniors, colleagues, and students have played an important role in the formation of the ideas developed. I take this opportunity to put on record the special debt that I owe to Professors A.K. Dalal, A.K. Mohanty, Albert Peptone, Anand C. Paranjpe, Anand Parkash, Arvind K. Sinha, A.K. Ramanujan, Ashis Nandy, Biranchi N. Puhan, B. Krishnamurti, D. Sinha, E.S.K. Ghosh, F.M. Sahu, Fathali Mogaddam, H.S. Asthana, John Shotter, J.B.P. Sinha, Janak Pandey, K. Ramkrishna Rao, Kenneth J. Gergen, L.B. Tripathi, Leela Krishnan, Mary Gergen, Martin L. Maehr, Matthijs Cornellison, Mckim Marriot, Michael F. Mascaolo, R.K. Naidu, R.C. Tripathi, Richard Shweder, Rom Harre, Stanley Kurtz, Sagar Sharma, S.C. Dube, S.C. Behar, S.N. Rai, and T.S. Saraswathi who on various occasions have provoked, encouraged, energized, and invited me to reflect. I have immensely benefited from them and express sincere thanks to them. Some of the essays took shape in collaboration with A.K. Mohanty, K.N. Tripathi, Suneet Varma and Uday Jain. I fondly remember the moments spent with them when the ideas were put together and hope that the dialogue shall continue.

The work was possible with the cooperation of my wife Madhuri who provided the necessary support and encouragement to work on the manuscript. My daughter-in-law Smita took care and enabled me to finish the work which was pending for some time. Manasi has helped in correcting the proof of the manuscript. Sri Ashok Mittal took keen interest in the production of the volume. I express my thanks to all these partners in this endeavour.

Delhi **Girishwar Misra**

Contents

1

Introduction

The contemporary human condition reveals that we have reached to a point where humanity feels empowered by an amazingly wide spectrum of achievements in telecommunication, media, means of recreation, transportation, health, and remarkable advances in diverse fields of knowledge. This, however, has also been accompanied by problems like massive population growth, environmental degradation, poverty, war, terrorism, violence, violations of human rights and problems of mental health. We are witnessing destruction of indigenous cultures, large scale migration and displacement of people, rapid westernisation, increase in cultural conflicts and clash of civilizations. Large scale and rapid social change and increasing gap in the distribution of wealth between the rich and the poor have created disjunctions and have led to the problems of adjustment on a global scale. The multiplicity in our contemporary problems demands developing a different kind of psychology which can be responsive to global and diverse life contexts, conditions and consequences. As Marsella (1998) argues, a global community psychology is the need of the hour.

It is now increasingly felt that the prevailing assumptions and methods as they operate in the mainstream have limited value and further growth of the discipline will depend largely upon the use of the wisdom and knowledge available in different cultures. The westernised global culture cannot be considered to hold true for each and every culture. As Wilpert (1998) has argued, is "the insight, and wisdom have been accumulated by many indigenous psychological traditions which psychology cannot afford to overlook. The challenge is of mining the hidden treasurers of these scarcely known psychologies." Wilpert hoped that psychology will be poised to successfully confront the present and future challenges which it faces. However, the preconditions for this are that psychology should, (1) widen the scope of societal issues, (2) bridge the mainstream and indigenous psychological approaches with curiosity and innovative spirit, and (3) use the international institutional networks available and offer to the world a united policy about the promise and potential of our discipline.

The International Scenario

The above set of realizations makes the contemporary scenario in psychology as one full of excitement. It is becoming multivocal and its boundaries are being redrawn. It has many facets and concerns. But all is not well and in order.

The prototypical mainstream psychology has maintained and nurtured the stance of natural science, and its methodological compulsions have resulted in the practice of a decontextualised science of human behaviour with the goals of manipulating, controlling, and predicting human affairs. The innocent or naive scientific curiosity which apparently seems convincing, legitimate, and deemed as a facilitator of human welfare; at a closer analysis, emerges to be one rooted in a peculiarly Western capitalist spirit. It is neither value-neutral nor globally empowering in its conduct. It has become a tool for maintaining inequality, oppression and injustice (see Fox & Prilleltensky, 1997). It thwarts promotion of human welfare and reinforces Western society's *status quo*. The ideology of individualism continues to dominate its practices and the same is celebrated through psychological technology.

The culturally indifferent conceptual structures and analyses of psychology make dialogue with people at large and with colleagues in other disciplines difficult and this makes the discipline an extremely alienated discipline. The use of psychological knowledge, therefore, becomes confined to the Western minded educated elite. The psychological technology has a clear urban bias and caters to the needs of the modernised industrial sector of society. It has ignored the concerns of the masses particularly in developing countries. It is, therefore, not surprising to note that psychology has largely remained absent from the arena of policy making. The climax of technology and instrumentation is reflected in the overuse and at times, misuse of psychological tests. This, however, has led to the proliferation of instrument dependent research which has questionable value with reference to generating knowledge or solving social problems.

The research scenario in psychology and partly in other disciplines too presents a condition that can be understood through the metaphor of market. In this context, the relation between knowledge producers and customers becomes important (Hurme, 1997). Knowledge is produced in relation to the nature of demand. The funding of research projects comes from various agencies which have priority areas and interests guided by market considerations. In fact, many companies and business houses have their own R & D departments. In this process, those ideas, concepts, and measures are developed and used which hold market value. In recent years, relatively greater emphasis has been laid on inculcating professionalism as against intellectual excellence. This is partly related to saturation in job opportunities in the academia. Professionalisation, it may be noted, is also in a way, a market driven trend.

A rethinking of metatheoretical and methodological assumptions underlying the science of psychology is currently under way. Developments in the philosophy of science have indicated that the very idea of science has undergone important changes and the historiography of sciences does not lend support to a simplistic notion of science which is still held in high esteem by psychologists in their research paradigms.

An analysis of science tends to show that it is a polymorphous activity. It has gone beyond the Newtonian paradigm. Appreciating the changes in the nature and conduct of science has far reaching implications for conducting psychology. In fact, creative responses to the inertia has resulted in many diverse kinds of voices (Maxwell, 1984). The deconstructions and reconstructions which have taken place in thinking about science offer new frameworks and create new grounds for practicing science. The developments in qualitative, context sensitive and participative methods of enquiry too are remarkable.

This post-positivist, new paradigm research is characterized by the concerns for understanding, meaning, interpretation, discourse, holism, particularism, cultural situatedness and subjectivity (Smith, Harre' & Langerhove 1995). A pluralistic, expanded and enlarged discipline of psychology is gradually taking shape. In this endeavour attention is being drawn to not only certain older traditions like phenomenology, symbolic interactionism and idiographic study but also to new theoretical perspectives such as social construction, social representation, feminism, discursive psychology, dialogical psychology, and narrative psychology. It is recognized that the social world consists of speech acts located in persons and conversations. Also, the behaviour of people has meaning for the people themselves and is mainly intentional. These realizations have started to change the mind set of psychologists.

There is also a clear move towards relating psychology to culture in significant ways. While it is difficult to deny the hidden agenda, a positive change showing give and take across cultures is noticeable. Progress in the fields of cross-cultural psychology, cultural psychology, ethno psychology and indigenous psychology have drawn attention to cultural factors in the shaping of methods, contents, and contexts of psychology. This movement has encouraged a process of cultural orientation the discipline of psychology (Gergen, Gulerce, Lock & Misra, 1996).

It is largely held that as a discipline, psychology is flourishing and we tend to take many things for granted. Questions about context, method, and its ontological and epistemological bases are usually not addressed. It is often forgotten that psychology is not a natural science as it is constituted largely by human social practices. The secularization of the world view and visibility of the success of the natural science model has not allowed us to appreciate human behaviour as 'meaningful', involving active agents with expectations and intentions and capable of communicating with other active agents.

The Indian Scene

Psychology in modern India is now eight decades old. During this period the discipline has grown in many ways as documented by the number of Ph.Ds. produced, number of publications and expansion of undergraduate and post-graduate teaching programmes across the country. Notwithstanding this proliferation of teaching-learning opportunities there has been growing discomfort amongst concerned colleagues that the discipline has not been able to achieve the status it deserves in the family of social sciences (Dalal, 1996; Sinha, 1995). In recent years, a critique of the progress of psychology in India has led to critical thinking and search for alternative paradigms. Psychology is moving towards a stage of reconstruction. This move, however, is visible in a small minority which is experimenting with new and innovative ideas and methods. The majority is still operating within a routine and replicative framework.

Reflecting on the present day reality of Indian psychology, one comes to realize that it has been a shifting reality that has undergone changes in both quantitative and qualitative aspects. At the national level we notice that psychology has been largely dominated by Western concepts, theories and methods. The Western orientation has unambiguously constituted the major academic force. The efforts toward developing indigenous psychological paradigms have been sporadic and very limited in scope. As a discipline, psychology like many other social sciences was an offshoot of philosophy. However, it opted to emulate physical sciences. It has helped to develop Western minded Indian psychologists who look for insights in concepts, theories and methods towards the Western world (Misra & Gergen, 1993a). Lacking in intellectual creativity, attempts were made to verify the Western notions uncritically. Critical reflexivity has been very limited, however, it has alienated the students of psychology from the body of indigenous knowledge, conceptual repertoire and knowhow available in the rich Indian tradition. The mindset developed due to colonial incursion was so powerful that although Western concepts were accepted and welcomed without scrutiny, indigenous concepts were banned and denied entry into the academic discourse (Misra & Gergen, 1993b).

Many psychologists still hold the view that India did not have any significant tradition of psychological thought. Our cultural subordination created a mindset which dichotomized tradition and modernity. Indian tradition now stands for our past and modernity represents the Euro-American present. Educated Indians are increasingly suffering from this kind of split in cultural consciousness. The living traditions of India do not attract many of us. A good Indian psychologist is often well versed with the nuances of Western psychology and its tools but has very little concern for indigenous knowledge.

In one of our studies we (see Misra, Prakash & Varma, 1999) tried to examine the disciplinary problems from the perspective of teachers and students. It was found that the psychological texts which provide intellectual support to maintain the universality of knowledge hardly provide any scope for indigenous wisdom. The psychological instruments also present the same picture. Research practices tend to be imitative even though there are dissociations between mainstream Euro/American concepts and the social and psychological reality encountered here. The Western conceptual structures often become problematic for Indian psychologists pursuing cross-disciplinary dialogue. Their urban-elite bias creates an isolation and alienation of the knowledge producers from the knowledge users. The inaccessibility of such knowledge for the potential users highlights a harsh fact that psychology has not done well as a social science though it promises to have the potential to improve the quality of life of the people. The imitative and replicative type of research has at times led to a derogatory construal of Indian culture and psyche.

Gradually evidence has accumulated which questions the universality of Western psychology. Psychologists have started taking notice of societal problems and have taken initiative to address various societal problems in the Indian context. It is now being felt that as a discipline, psychology can no longer afford to continue with the image of a pure scientific enterprise committed merely to conceptual and methodological advances. It needs to link itself with life and become meaningful and fruitful for individual and society. To this end, psychology has to mould its working substantially in its contents as well as a mode of delivery. It requires a reorientation so as to attend to the demands of diverse sociocultural spheres. As in his autobiographical account entitled *'In Search of My Brahman'* Jai B.P. Sinha (1997) rightly maintains: "*Neither our approaches, nor our methods, nor our priorities match western concepts and theories which are at best only partly applicable in the Indian setting. I do not have the need to carry the cross of someone's theory or approach, nor do I have resources to try mine in some other cultures. My hunch is that even if I try, I will be imposing my emic under the delusion of realising an etic. May be the best I can do is to try to be a cultural rather than a cross-cultural psychologist. May be the route to a universal science of psychology is through many indigenous psychologies forming an overarching pattern. We all have our own islands—psycho-historically and culturally determined. All we can do is to explore our own island and build bridges to communicate and share. These crisscross bridges will some day assume patterns which we may call the facets of universal psychology…*" (p. 83)

The Question of Relevance

The question of relevance of psychology is a relative question. It demands specification of a context. Thus relevance may refer to individuals, groups or the society as a whole. Also, it requires specifying the domain in which relevance is sought. Thus relevance may be for ideology or for behaviour. In any analysis, however, the question of relevance has to be decided by taking into account, the target and essentially involves choice of a position. It is obvious that the position which we have to take must suit the target. Thus making something relevant will go against *status quo* and requires a move in favour of an altered position. In order to be relevant, therefore, psychology has to change its value neutral posture.

Relevance is dynamically located in time and space. The contours of any society are always in a process of flux. The situational demands change from time to time and, therefore, the criteria for relevance also change. Relevance involves change and transformation and brings us in a state of disequilibrium. It may not be a very systematic, logical and sequential process. It demands attention to heterogeneity, non-linearity, and openness. Being relevant requires a creative mindset which has the intellectual courage to experiment and test new ideas. Of course, such innovations or breakthroughs may not be assuring.

In order to make psychology relevant it is necessary to address the state of affairs in a dispassionate manner. 'We have to ask ourselves the questions why are we doing psychology'? and 'where are we located'? The neutral stand has already proved insufficient and instead of a deficiency fulfilling model we have to creatively contribute to the formulation of social values. Psychology also needs to become sensitive enough to the actualities of contemporary life. Making teaching-learning more involved and skill oriented would go a long way in repositioning the discipline. Also, there is a need to have a holistic perspective which is participatory in nature. In view of the current state of affairs, psychology can be nothing but a multivocal enterprise.

With increase in specialization, the various sub-specialities developed links with other disciplines outside the field of psychology. The relationships between different sub-areas or specialities in psychology became tenuous and no effort was made for their meaningful integration. This fragmentation is continuing. We have not been able to reach a unified view of human nature yet. For the viability of psychology as a discipline, we depend on the way it relates to the changing contexts of life. We need to bring critical issues into focus and offer analyses and explanations that indicate alternative policy options.

In this context, the question of societal development needs serious attention, which is resulting in poverty and social disharmony. We need to question who is deciding and for whom. In a non-participatory process of development, the elite is the decision maker. A liberatory or emancipatory psychology will need mass consentisation, organization for participative decision making and resistance to the vested interests.

Thus there is an urgent need to critically think whether predetermined categories of description, measurement and analyses are adequate enough for the comprehension of alien cultures and societies. The consequences of western theories and methods for the native categories, meanings and thoughts cannot be brushed aside. Dominant paradigms should not be allowed to distort the social and cultural reality and neglect their meanings, instead of continuing as prisoners of traditional thinking dominated by Western models, we need to respond creatively using new categories. Having an empathic attitude, we need to develop a psychology which is not merely modulative and reactive but one which has a generative vision that may promote well-being of the individual as well as the collectivity. The chapters that follow attend to the dilemas, struggles and developments involved in following such a move in the discipline of psychology.

2

Potentials of Indian Thought

As noted in the preceding chapter, recent developments in psychology are beginning to question the fundamentals of this scientific enterprise. The last hundred years of psychological science have led us to different realms of our existence and have definitely influenced our thinking about ourselves and the surrounding environment. This kind of psychology which largely grew in the West was an outcome of the demands of a specific socio-cultural and ecological milieu (Buss, 1975; Popper, 1967). But the theories, however, were proposed for universal application. The gap in applying these theories to different socio-cultural groups has become obvious in cross-cultural investigations. The properties and expressions of basic processes like perception, learning, and thinking are found to be linked with cultural boundaries.

A critical scrutiny of Western psychology's overemphasis on experimental method as the most appropriate method to approach reality has not been able to fulfil the expectations. The discrepancies between scientific formulation and reality constrain our contemporary knowledge of psychological processes and restrict their generalizability. The experimental approach, being less sensitive to certain important aspects of reality, provides a distorted picture of reality. It reflects, what Koch (1981) has termed the *syndrome of ameaningful thought*. Koch draws attention to the limits set upon the attainable knowledge set by the subject matter of psychological studies that make the natural science model inapplicable to psychology. The analysis of experimental method clearly evidences that its assumptions about independent variable (Anandlakshmi,

1974), control and prediction are fraught with limitations (Hendrick, 1977; Jain & Misra, 1985). The theories of human behaviour that evolved through experimental studies, have largely been inconclusive. The sophisticated experiments on learning and memory carried out in the last 100 years have provided knowledge of memory which a 10 year old holds (Neisser, 1978). They tell very little about the properties of memory as experienced in natural life circumstances. Under the influence of evolutionism psychologists devoted their energies to the study of animal behaviour and aimed at generating principles of behaviour. However, the experiments on long delay learning, operant conditioning of autonomic responses, comparative cognition, and semantic memory have questioned the basic premises of our most sophisticated scientific investigations.

Recent years have evinced a revival of interest in the study of 'self' among psychologists. In fact, 'self' has been welcomed back into the realm of legitimate constructs within almost every branch of psychology. This interest is in both the aspects of self namely self as actor or subject (I) and self as on object of one's knowledge (Me). More and more attention is being paid to the processes through which knowledge of the self contributes to regulation of our behaviour. Self is considered as a cognitive structure or inference which is empirically derived. The individual is increasingly being viewed as an active agent. Its cognitions about the world and itself are found to significantly mediate its behaviour. Self is being viewed as a major concept in the theories of self efficacy, self consciousness, self control, and self regulation. These developments suggest a new language which has heavy emphasis not on the objective reality but on the way reality is subjectively constructed by the person.

The studies of consciousness and related issues are also indicating that our assumptions about brain as a limited information processing system is untenable. If brain has to be considered as a machine it should be treated as a biological machine having the characteristics of maturation and learning. Learning always modifies the brain and, therefore, brain has tremendous possibilities. The limits of cognition (consciousness) are set by those things which we have not learned. Also, the episodic model capable of dealing with discrete behaviours or responses is too simplistic to deal effectively with the stream of behaviour (Misra, 1984). The substantive limitations and conflicts faced by psychologists in the West have led to a search for alternatives and new conceptualizations are emerging. For instance, control is being considered as the mindful process of mastering (Langer, 1983). Actual behaviour is viewed as a continuous process of multidirectional interaction between the individual and the situation (Endler & Magnusson, 1976). The relation between the object of study and method of research is being questioned (Mamali, 1982) and attempts are being made to develop a participatory psychology (Stringer, 1982). The language and content of Western psychology is changing and the trend indicates that psychology is regaining its mind.

Academic Psychology in India

Western psychology as a discipline in India is not very old. It has less than a century long history. It was borrowed from the same German tradition from which American psychology emerged. Perhaps the difference lies in the fact that the Americans borrowed the experimental method and gave it American substance while Indian psychologists imported wholesale (Sinha, 1984). Psychology, therefore, grew as a discipline which lacked a deep or organic relationship with the culture, society, and social reality of India. It dealt with the problems borrowed from Western theories through the methods which were prevalent in Western psychology. Indian universities base their teaching on American and British textbooks and hardly refer to the work done by Indians in India. The growth of psychology in India is largely a post-independence phenomenon. However, a colonial spirit prevailed and a fascination with 'Foreign' made Western psychologists as the significant others continued. Our outward gaze made us ambivalent to indigenous scholarship and defeated the process of serious internal dialogue among psychologists in India. The uncritical recipient culture which grew in India led to non-reflective activity manifested in replicative, arbitrary and adhoc research (Nandy, 1974; Sinha 1983, 1984, 1986).

The borrowing of conceptual models and methods led Indian psychologists to view and respond to the Indian social reality as given in the American books and not as we experience in India. The studies conducted in this context, therefore, remained unrecognized and unused by the Indian psychologists themselves. As pointed out by Sinha (1984, 1985), the recipient culture of psychology has been a major cause for this state of affairs. In addition, the microsocial orientation of theories, their development in relatively stable Western societies, methodological constraints and lack of interdisciplinary perspective have also been responsible for the limited impact of psychology.

In recent past some interest has been shown towards the application of psychology to problems of social change and national development. A minority of psychologists are trying to understand social problems in Indian setting in a realistic manner from a culturally appropriate angle. The awareness of our limitations and a desire to establish the identity of Indian psychology is clearly evident in academically active psychologists. However, a majority of the students of psychology still hesitate to talk about the Indian heritage and its study has not been able to receive the attention of Indian psychologists (Sharma, 1981; Sinha, 1984). Even the four surveys of psychology sponsored by the ICSSR have not found it fit for inclusion. This attitude further attests our hesitation and reluctance to accept the study of Indian psychological perspectives as a legitimate field of psychology. Our hesitation is due to our general belief that the traditional Indian thought is irrelevant to the demands of the time and that they do not help in the process of achieving scientific principles of psychology.

A critical appraisal of the situation suggests that traditional thinking or any model is limited in its applications. The range of problems that can be solved are specifically linked to the model. The strength of traditional thought is evident in its internalization by the masses and its acceptance by them as guiding principles in their lives. Attempts to study the nature and significance of traditional Indian thought may prove useful in linking psychological knowledge with real life settings. In recent years researchers have been realizing a need for translating psychological knowledge into the concepts and language of everyday experience so that the knowledge may be applied (Stringer, 1982). In addition, common sense is also being considered as an important source of psychological knowledge (Fletcher, 1984). These initiatives will bridge the gap between the expert's and layman's views of the social world and enhance the quality of the knowledge generated.

A universal psychology can be possible only if we have an understanding of specific cultures. Intellectual exercise to study Indian thought, therefore, is rewarding in two ways. First, it would allow to view reality from a culturally congruent perspective. Second, it would enrich psychological concepts and theories. Some of the important areas in which indigenous Indian thought can contribute and extend the Western view of psychological processes include self, states of consciousness, philosophy of mind, yogic methods and ways of improving the human condition. The need to integrate eastern and western traditions and evolving a non-Western perspective has been indicated (Sinha, 1965, 1981). Initial work to organize the relevant materials was done by Akhilanand (1948), Ramchandra Rao (1962), Gupta (1977), Kakar (1978), Krishnan (1974), and Singh (1971). Some western scholars have also shown interest in Indian psychology (Hall & Lindzey, 1978; Murphy & Murphy, 1968). The work of Prabhu (1963), Safaya (1976), and Kulkarni (1984) have provided solid ground for establishing Indian psychology. Paranjpe's (1984, 1998) Volumes *Theoretical Psychology* and *Self and Identity in Modern Psychology and Indian Thought* are excellent works which establish beyond doubt that the study of Indian psychology is academically rewarding and relevant to contemporary psychology.

Paranjpe (1984, 1998) has shown that traditional Indian thought includes important theoretical work on meditation, self, identity, yoga, counselling, and mind. It provides a world view which has significance in understanding human relations and the realization of potentials. Theoretical emphasis on the structures and processes relating to different levels of personality are available. However, the relationship between religion and science as found in West is not applicable to India. In the Indian context of Adhyatm, there is no conflict between science and religion. This conflict in the West has resulted in the avoidance towards religion. In contrast, Indian science developed in close relationship with religion (*dharma*). This led to the rejection of a clear division between mind and matter. Cognition was the main goal of scientific understanding. Contemporary psychology is also going through a cognitive revolution. Search for understanding is being viewed as the basic spirit of action. This change is quite congruent with the spirit of Indian thought. Let us examine some of its features.

A Different Perspective on Reality

The study of Indian thought is relevant to contemporary psychology because it views reality from a perspective which has possibilities of suggesting new paradigms and strategies to study psychological phenomena. The Western psychology defined its goal as control and prediction of behaviour while self-realization became the goal of science in Indian thought. The Indian view assumes that the universe is lawful and there is continuity within the entire animal kingdom. The difference between material and moral as found in the West is not applicable to Indian thought. Similarly man's relationship with nature is conceptualized in a different way. The Western view emphasizes a conflict between man and nature because man is placed over nature. Such a conflict is absent in Indian tradition in which man and nature are interdependent and share coexistance. The Christian model views a cause and effect relationship between man and universe, while Indian view assumes an inherent interrelatedness. Vedic literature shows that the relationship between man and man, and man and universe is of interdependence. In Indian thought, as V.N. Misra (1984) has remarked, man does not stand on the top of universe nor the rest of existence is subservient to him. There is an inherent correspondence of macrocosm and microcosm. The man in is operating on two planes, one operating along the dimension of time and space while other transient time through its projection in its progeny and transcends space through its identity not only with the nature around but all the beings. The Indian view does not reject history but transcends it. Western thought views man as historically situated. Therefore, nature or physical energy is considered different from man leading to competition between the two. Man tries to master and control nature. In contrast, Indian world view perceives a continuity between man and nature. Therefore, there is coexistence and congruence between man and God. Man is not the centre of thought. The interrelationship among various aspects of reality is important in which nobody is consumer in an absolute manner.

Paranjpe (1984, 1998) has drawn our attention to the differences in the conceptualization of human nature in Indian and Western thought. He suggests that the Western view is predominantly hedonistic while Indian concept of human nature deemphasizes the concept of power seeking man. An individual enjoys pleasure and suffers from pain. Here, self is emphasized for self realization and not for self centeredness. Western theories also take a different stand on rationality of human beings. Some consider human nature as rational and others view it as irrational. Indian system does not emphasize either aspects of man. The rational and irrational behaviours are temporary manifestations of *gunas*. So far as human condition is concerned there is superficial similarity between eastern and western views about suffering. However, the experience of suffering in Indian condition is related to the quest for knowledge. The ideal human condition is not important in western psychology but Indian tradition is concerned with highest and most lasting form of happiness. Some concepts proposed to describe this state are Sreyas, Kaivalya and Jivanmukta. The concept of Sat, Cit and Anand also refers to this state. These concepts suggest the centrality of self- realization.

The analysis of social transactions has also been attempted in different ways in the Indian and Western contexts. In a systematic examination of these differences Marriot (1976) draws our attention to the facts of identity of actor and action and divisibility of the person. He argues that "the Indian view does not presume the separability of actors from actions. It presents monistic cognitions of reality. What goes on between actors are the same connected processes of mixing and separation that go on within actors. Actors' particular natures are thought to be results as well as causes of their particular actions (*Karma*). Varied codes of action or codes for conduct

(*Dharma*) are thought to be naturally embodied in actors and otherwise substantiated in the flow of things that pass among actors" (pp. 109-110). Persons-single actors, are not thought to be individuals i.e., indivisible or bounded units. Persons are dividuals or composites of substance-codes that they take in. "To exist, dividual persons absorb heterogeneous material influences. They must also give out from themselves particles of their own coded substances — essences, residues, or other active influences — that may then reproduce in other something of the nature of the persons in whom they have originated" (Mariott, 1976, p. 111). However, most of the Western scholars have preferred the practice of postulating Indian dichotomies equivalent to the Europian dualities.

The human transactions thus conceived have significance for psychological theorization. As Prabhu (1963) states "they are very systematic and are parts of the deliberate planning, organization and system, a whole thread of an ideology running through a consistent social philosophy underlying all the institutions and knitting them together into a social pattern as social fabric" (p. 342). This pattern did not have "any opposition between social good and social ends on the one hand, and of individual good and individual ends on the other" (Prabhu, 1963, p. 357).

Thus, it is evident that the Indian perspective in reality, approaches the psychological and social processes in ways which are different from the Western view. A detailed study of this field would prove fruitful in reconstructing our understanding of psychological processes from a wholistic perspective. In addition, it would be useful to those who are interested in developing universal psychology dealing with ethics. Paranjpe (1984) emphasized that there is a need to construct a broad common theoretical framework based on a set of common ideas found in many theories across cultures. This should be followed by adding other closely equivalent concepts and roughly equivalent and complementary ideas.

Paranjpe (1984) has illustrated this mode of analysis of comparing and reconstructing the Western model and Samkhya model. He refers to the current systems theory which utilizes the concept of self-regulation since the system monitors its own monitoring. He also considers the thought as emergent function which influences body. Information processing occurs along a continuum consisting of mechanical, electrical, biological and psychological methods. Referring to the work of Popper and Ecclas (1977) on cosmic evolutionary stages Paranjpe (1984) identified three worlds of existence. World I consists of physical, chemical and biological entities. World II refers to subjective experiences and World III consists of the products of mind. A similar approach is observed in the treatment of 'Prakriti' in Samkhya theory as described in *Tamas, Rajas* and *Satva* gunas. Both of the views assume interactions among the three components. However, the concept of consciousness in West is viewed as all or none phenomenon while Samkhya Yoga theory conceives of levels or grades of consciousness. Paranjpe (1984) also draws our attention towards the similarities between systems theory and Yoga. Both assume that the whole has emergent properties that are irreducible to parts. It is a method of establishing a total voluntary control of the cybernetic mechanism which the mind-brain system has at its disposal. Yoga as a method of self-regulation involves voluntary regulation of thought processes which may ultimately lead to no thought zone of consciousness. Paranjpe (1984) has also referred to the similarity between phenomenological cognitive approach and Vedantic approaches.

Thus, we find that the study of the elements of psychological thinking in Indian thought systems is relevant to contemporary psychology for generating hypotheses, providing a new perspective on psychological processes, understanding human relations, and changing the individual in a positive direction. It will help studying Indian problems from an insider's perspective. This, however, requires addressing the question of scientific concerns of psychology as Indian thought has often been treated as nonscientific.

3

Problems with an Acultural Positivist Psychology*

One of the chief concerns and commitments of psychology has been to gain scientific understanding of the behavioural phenomena and attaining the status of science within the framework of physical sciences subscribing to positivist philosophy. While emphasizing determinism and operationism this framework projected, controlled and objective observation of behaviour as the most effective method to formulate universal laws of behaviour. The dream was to delineate such laws, as are proposed in physical sciences, and to develop advanced technology to control and predict human behaviour. This approach was primarily method-centered and treated method as a critical factor in determining the content and quality of knowledge obtained. A close perusal of the various fields of psychology reveals that the domain of experimental psychology partially fulfils the stringent requirements of scientific method. Attempts to observe relevant phenomena in isolation or under controlled conditions in other fields such as personality, social psychology and human development, clearly indicate that the experimental method is insufficient. It is chiefly because not only the required degree of control is not possible but also because the experimental approach tends to distort the reality under investigation and misplaces the research priorities.

In recent years psychologists have become increasingly aware of the limitations in scientific study of psychological

* The ideas developed in this chapter emerged in collaboration with Prof. Uday Jain.

issues. It is being widely felt that the models of physical sciences which propagate development of abstract universal laws assumed to have applicability without the constraints of time and space are inadequate to explain and predict the complexities of human behaviour. Also, there is growing consensus that knowledge in general is a socio-historical product. These two realizations combined with the awareness of the inherent properties of the subject-matter of psychology, which make it less appropriate for such an inquiry leads to the suspicion that a very high level of predictability of behavioural phenomena is inconceivable. It is considered that we need a new paradigm of science and a new model of enquiry. The inevitable consequences of too much preoccupation with operationism was that the knowledge which at best constitutes one of the perspectives on reality came to be treated as the representation of reality. In actual practice theories often present different perspectives on a given aspect of reality and, in turn, determine the extent of our accessibility to different aspects of reality.

As Gergen (1985) has articulated, "the process of understanding is not automatically driven by the forces of nature, but is the result of an active, cooperative enterprise of persons in relationship" (p. 267). He proposes that 'the degree to which a given form of understanding prevails or is sustained across time is not fundamentally dependent on the empirical validity of the perspective in question, but on the vicissitudes of social processes (e.g., communication, negotiation, conflict, rhetoric)' (Gergen, 1985, p. 268). This view is contrary to the empiricist conception of scientific knowledge which considers knowledge as an internal representation of the state of the nature, which has formed the rationale for research in psychology. A close scrutiny of the prevailing state of affairs makes it clear that there are many problems with the empiricist notion of scientific understanding when applied to the human phenomena.

In this context the role of commitment to the evolutionary theory of behaviour which assumes a continuity across species and favours reductionistic explanations of the phenomena under study also needs scrutiny. Following the premises of reductionism it was believed that gradual understanding of simple forms of behaviour will eventually enable us with the understanding of complex forms of behaviour which essentially differ from the former in quantitative terms. This assumption has been the implicit or explicit guide for majority of the studies done in the last hundred years of psychological enquiry. This situation had important implications for the growth of psychology as an independent and unique science. In order to establish it as a distinctive science, psychology has to move beyond showing by analogy, how persons are like animals or things. Rather, it should strive to develop principles uniquely applicable to persons.

Human mind has the capacity of self-reflection and action. The ideational dealings at human level, make the individual partly free from the limitations of time and space. As Nuttin (1984) has indicated, ideational dealings are loose, flexible, reversible, and make it possible to handle the things in representational form. To give an illustration we can think of the significance of future orientation. Behaviours in reality are not limited to the present; rather they are almost always extended to the future. The future is of course represented in the present but it has its own existence and has bearing on the present. Granting self-reflection, demands acceptance of 'natural observers' as a active and concrete dimension of our enquiry (Mamali, 1982). The personal (and direct) experience should be granted authentic status as a source of data. In the name of 'objectivity' we mostly rely on indirect and second hand data to base our generalizations.

Following the natural science paradigm, the emphasis is on one set of events as antecedents of another set of events. In this process we desire to eliminate all other variables except one, the presumed cause, so that we may observe under controlled conditions some isolated parts of the situation-person complex. Contrary to this, our generalizations are often assumed for the total reality (the whole person across situations). This kind of over-simplification often frustrates those desiring for accurate prediction and control.

Psychological endeavours to understand human behaviour are confined to specific behaviour-episode (*e.g.* lever pressing or recalling a word), which represents a very tiny segment of behaviour in the real life. Real behaviour, as Atkinson and Birch (1978) have asserted, involves a stream and has continuity. Psychology has largely been concerned with episodic behaviours. The dynamic nature of the flow of behaviour cannot be approached or appreciated following the simple episodic model with a focus on discrete behaviours.

The most revealing imbalance between concepts and methodclogy is found in the field of social psychology. A number of social psychologists have questioned the social nature of social psychology (Sampson, 1978) and have argued that experimental approach distorts the social reality. The theories of social psychology are developed in the form of general psychological theories while it is apparent that social psychology is related to changing social reality. As Gergen (1973) had pointed out, social psychology is nothing more than history of contemporary North America. Social phenomena are inherently more prone to the influences of immediate circumstances and contemporaneous factors. This has escaped the notice of most of the theorists while explaining phenomena across situations, cultures and time periods.

As Sigmund Koch (1981) contends, psychology has become misconceived either as a science or as a kind of coherent discipline devoted to the empirical study of human beings. He has noted that the nature of subject-matter of psychological studies is such that limits are set upon the knowledge that can be attained. According to his assessment, psychology cannot function with the natural science mode. It requires a mode of inquiry rather more like those of the humanities than the sciences.

Cross-Cultural Psychology: A Possibility

Cross-cultural psychology is the empirical study of the members of various cultural groups who are known to have had significant and identifiable experiences leading to predictable and theoretically important similarities and differences in behaviour (Brislin, Lonner, & Thorndike, 1973). It has contributed to the growth of psychology in a number of ways. It has increased the range of variables, helped in unconfounding variables, and has initiated examination of the context in which behaviours naturally occur. However, a close scrutiny of the field evinces that it too has many limitations. Triandis (1980) has noted that the meaning of culture is not clear and there is a need to unpackage it through implying ecological as well as self-report measures. The analysis of personal variables should be done within the etic, emic, and ideographic framework. However, the analysis of person-culture interactions still remains unclear.

Most of the cross-cultural researchers utilize tools, tests, and instruments which are imported from a different culture and the results thus obtained are used in the form of generalizations called *etics* which are supposed to characterize universalities. This practice has produced *pseudo etics or imposed etics*. We need studies following "the emic plus etic methodology".

Use of statistical techniques in analyzing cross-cultural data further complicates the problem of interpretation and explanation of behaviours across cultures. Statistical reasoning presupposes a specific distribution of the phenomenon under study. This assumed nature guides the decision for analysis. Such formulations may hold true for many physical or physiological events but their applicability to psychological phenomena is questionable. Even Hume's contention that scientific laws are statements of constant conjunction between events has been rejected by many scientists (Moricas, 1982). Therefore, applying statistical models to the study of cross-cultural behavioural data would be inadequate because different cultures may follow different laws about different phenomena.

The theoretical models and research based on them are by and large Euro-American products. It is not clear whether they provide a universal account of the phenomenon. As Triandis (1980) stated, 'we do not know whether our theories provide universal generalizations. They infact ignore the significant variables and phenomena that occur in non-Euro American cultures'. Looking at the current developments it may be said that cross cultural psychology has helped in increasing the range of variables to be studied and has provided a way to examine the context in which behaviours occur and has proposed some new hypotheses. However, much has yet to be done to uncover the meanings of concepts across cultures. Imposition and importation of concepts from outside a given culture has to be replaced by identifying the core concepts and their culture specific meanings. It is, therefore, widely felt that we need indigenous cultural psychologies for attaining the goals of universal psychology.

Needs for Indigenous Psychology

The behaviour of individuals in any particular culture is guided and controlled by the concepts they internalize in that culture. Therefore one possible way to formulate a theory of behaviour following an emic approach is to identify the concepts in a particular culture and then to find out the suitable methodology to investigate them.

In order to illustrate the issue we may consider certain features of the Indian tradition. Many of its concepts still have central place in the day-to-day life of the people and they do guide their behaviours. An analysis of these concepts may provide insights into the functioning of behaviour. It may enable us to develop a culturally more meaningful and relevant psychology. As Harre and Secord (1972) have emphasized, the conceptual system embedded in ordinary language should provide the basis for the concepts employed in a realistic psychology, and should serve as a model for other logical connections and new concepts introduced by psychologists. To know these concepts, a less structured and an unobtrusive method has to be employed (Bond, 1983). Perhaps direct reports of feelings, plans, intentions, reasons, and beliefs would go a long way in understanding behaviours in specific socio-cultural settings. In an innovative work of significant concepts in the laymen, Sharma (1988) has tried to see the meanings of various philosophical concepts by the lay people. The findings have shown that understanding of these concepts varies as a function of subcultural variables like religion, education and socio-economic status. Inspite of variations across groups they do provide a value-belief structure of Indians which forms a proximal determinant of day-to-day behaviours including inter-personal relationships, social perception, and attribution of causality.

In the Indian context efforts have been made to interpret various concepts such as *Atman, Dharama, Brahman, Purusa, Chitta and Manas* etc. However, at large such efforts either try to seek similarity of meaning with modern psychological treatments or try to supply another version of classical Indian texts. Some psychologists accept the need of culturally specific concepts to be studied but borrowing methods from the West in their opinion, has no harm. This seems to be a self-contradictory stance, for such methodology may set limits in the sense that operational definition of the concepts may distort their original meaning.

Heider's (1958) emphasis on studying the 'naïve' psychology or the study of layman's understanding of the causes of events and interpersonal behaviours is another approach. The emphasis here is on knowing under what conditions the attribution is made to certain factors. In attribution process the characteristics of the attributor, the attributee, and the situation interact in such a fashion that pure logical rational attributions are rarely made (Jain & Misra, 1986). In most of the cases various cognitive or motivational biases distort the objective inferences. Perhaps in social sciences such biases are rules rather than exceptions. Applying the Western attribution model (e.g. Kelley, 1967) to understand the perceived causes of behaviour or events in Indian setting, particularly in Indian villages, would necessarily lead to false conclusions where the causal categories such as 'Karma Samskar' 'Caste' 'Kul' 'Evil' 'Powers' 'Rinas' 'Bhagya' 'Pap' 'Poorv Janm', etc. are in common usage. This argument gets support from the studies on cultural differences in cognition. In fact, one cannot say that a particular cognitive process is absent in one culture and present in the other (Cole, Gray, Glick & Sharp, 1971), because the cognitive processes cannot be understood without the consideration of a situation to which they are applied. Thus, a situation-specific methodology is warranted.

It seems that the methods of investigating psychological problems should evolve within the framework of socio-cultural system in which the investigation is situated. The method should be sensitive enough to the behavioural functioning of participants. The studies have clearly indicated that people differ in the way they respond to items, with a number of categories (Jain, 1979). The responses also vary in relation to the contextual variables. For example, Sinha (1982) found that villagers respond better to interviewing in a group than an individual situation. The indiscriminate use of measures without taking into account the repertoire or responses of a given group may lead to disastrous results and misleading generalizations. In this context some innovations have been made. For instance, Sinha (1969) has used grain sorting as a measure of level of aspiration and behavioural ratings and happy life projections as indicators of motivation in villagers. Attention to culture-specific methodological problems has been given by many researchers. Tripathi (1982) has analyzed the problems of measuring deprivation in rural settings. He has noted that eliciting information from villagers is dependent upon *dress, deportment,* and *diction* of the investigators.

The significance of cultural context in the measurement of psychological attributes has significance from a theoretical angle also. As per the requirements of scientific rigor theoretical concepts are operationalized in terms of certain behavioural or response measures. Therefore, the behavioural characteristics guide us in the measurement of concepts. The framework developed in one culture may provide one scale of concept appropriate in that culture but the same becomes biased when put to use in another culture. The case of research on need for achievement (*n*-Ach) in India and other developing countries clearly shows this problem. By and large the poor economic development in these countries was atttributed to the low degree of *n*-Ach. The psychologists influenced by this suggestion endeavoured to demonstrate and confirm the proposed linkages in theory without looking into the Indian context. Recent studies on achievement motivation indicate that the competitive individualistic notion of achievement propounded by McClelland (1961) fails to recognize other possible modes of achievement (Sinha & Pandey, 1970). As Misra and Agrawal (1985) found the concept of achievement involves non-competitive goals and social means also. Recent studies in the field of attribution also suggest that the dimensionality of causal categories as suggested by Weiner (1979) is not applicable to several causes (e.g., fortune God, facilities). They also found that the division between "internal" and "external" was not clearly demarcated with respect to several causes as it is found in the Western cultures (Dalal, Singh & Misra, 1988). These results suggest that the conceptual foundations of methodology should be linked to the socio-cultural matrix. It appears that there is a need to look into the problem of matching methods with the concepts in the context of specific cultural settings. This will enhance our understanding of the concepts as well as provide useful information applicable to the cultural setting. Also, this strategy would generate findings which would facilitate the growth of indigenous psychology.

Unfortunately, owing to a colonial background, and projection of Euro-American psychology as 'universal' Indian psychologists have been trapped in a peculiar state of affairs. Owing to the Western education and pseudo modernism, sections of the educated Indian elite have developed disrespect for traditional learning, indigenous concepts and their own socio-cultural roots. A disoriented teaching-learning process has resulted in misplaced identities and imbalanced priorities. It is only recently that some social scientists have become aware of the situation and have made efforts of rediscovering indigenous concepts and formulating theories. The reorientation needed in our approach to handle this situation demands change in our attitude towards indigenous mode of thought and to overcome the limitations of scientism.

Use of indigenous concepts would go a long way in psychological knowledge with real life setting. Several lines of development in contemporary psychology show a positive intellective climate congenial for the growth of indigenous psychologies. This is reflected in the increased emphasis on translating psychological knowledge into the concepts and language of everyday experience (Stringer, 1982), critically examining the relevance and implications of empirically based knowledge (Kukla, 1989; Prelleltensky, 1989) considering common sense as an important source of knowledge (Fletcher, 1984). The indigenous psychological approach is rewarding for the development of the discipline in general, as well as to confront the social issues and problems in the cultural context. The interface of psychology and culture needs more serious attention in the post industrial world.

4

Understanding the Culture-Psychology Interface

In its evolution in this century, psychology as a discipline has almost exclusively focused on the individual. It is indeed paradoxical and interesting that this obsessive individual orientation has prevailed despite the early recognition of the close linkages between culture and psychological processes by the founding fathers of psychology. For instance Wundt's work on folk psychology explicitly emphasised those linkages. Questions may now be raised about the causes of this paradoxical professional amnesia, that has banished the inseparable cultural context from the minds of psychologists while theorizing about psychological processes.

Perhaps, the empiricist-positivist ideology that laid stress on the "value neutrality" of scientific endeavour and asserted with equal vigour the "universality" of constructs, was responsible for the (mis)appropriation of the methods and content of psychological investigation. Whatever did not conform to this inflexible notion of science was considered unworthy of the attention of scientists. As culture could not be easily reduced to the status of yet another "variable" influencing behaviour, psychologists substituted it by individual level intrapsychic concepts. These constructs, believed to be vested with concrete existence, were invoked to explain other psychological variables. In due course of time, the vision of psychologists came to be narrowly limited to the intra-individual processes encapsulated within the self-contained individual. Thus, culture was manoeuvred out and marginalised by the mainstream psychology.

The emergence of information theory during the post-Second World War era convinced psychologists that they were approaching a breakthrough. The cognitive revolution appeared remarkable, and influenced almost all areas. Thus, social psychology vigorously used information processing model. Within this framework, the concern, with the individual mind's inner workings was crucial and the study of social cognition became nothing more than a special case of information processing. Such context-free, intrapsychic and ahistorical analyses ignored culture's formative contributions.

A microindividual focused social psychology flourished, that could hardly make our social world intelligible. Further, the acquisition of cognitive structures and the relationship between cognition and action could not be explained satisfactorily. But this perspective prevailed in socialising students of psychology through American textbooks of social psychology. Being confined to the analysis of individual's behaviour in relation to social stimuli this approach remained reductionist in its explanation and denied the analyses of social qualities of human experience.

In the last few decades, social enquiry has been engaged in a re-examination of its purpose and methods. Conversation, participation in discourse and enhancing sensibilities are valued. The communicative and hermeneutic rationality is achieving greater significance than the search for an ultimate, literal truth. It is now being realised that constructive social efforts underlie social cognition. Interaction, negotiation and inference are continuously being used in such efforts. It is increasingly realized that practical activities, linguistic codes and social conventions constitute our social reality.

Human beings have a special role as they are simultaneously objects of cognition as well as cognising subjects. Understanding, therefore, involves constructive efforts and cognition is inherently social. Despite, the ideas of "collective representation", "thinking society" and the interpretive-interactive tradition in social science, psychological perspectives on social cognition have long remained unconcerned with social and cultural influences. Psychologists expressed their reluctance to accept cultural factors as they felt that their identity as hardcore scientists was at stake.

In recent years, not only have the epistemic foundations of positivist notions of knowledge been questioned but the need for indigenous cultural psychologies has been voiced at various forums as well. Researchers are now articulating the social origins of mind and cognition and cultural practices are seen as playing a constructive role in forming cognitions. The significance of cultural resources and social context is being recognised. Human participation in culture and the use of the cultural toolkit, as Bruner (1990) claims, works as a prosthetic device to amplify human capacities. Cultures often provide the worldviews through which people look at the universe. This implied that Western psychological concepts and theories are the products of a particular set of socio-historical conditions and are not universal. This is being increasingly realised in cultural psychological studies that go beyond "culture as a treatment view" and try to capture the reality of intentional worlds in which meaning and interpretation are critical. In this tradition cultural differentiation in human nature is related to the view of the person as a semiotic being for whom the historically acquired meaning of a situation or stimulus acts as the major constraint on response. The different situations and events may elicit diverse responses which appear rational in different local settings. In this perspective, cultural meaning systems, which are in certain ways local, become of central concerns.

Thus, a culturally informed perspective is emerging which looks at human action as culturally coded activity. European social psychological studies of socially shared reality and attempts to analyze narratives, argumentation and discourses are pointers to a new intellectual climate. Recognising the legitimacy of relative realities and pluralistic world-views indicates a decentering in our metatheoretic orientation. As culture is beginning to be seen as socially shared information coded in symbols, it has become a central concern for the students of cognition. Symbols become intelligible only when they are socially used and practised. This changing discourse focusing on a culturally-oriented social science is both challenging and exciting.

Historically the linkages between culture and psychology (mind) have fascinated the students of human sciences since time immemorable. In recent years these sciences are increasingly showing greater concern for culture in its diverse forms as a relevant context as well as a significant facet of human psyche. However, there is no consensus about the meaning of culture among the scholars of different persuasions. In this context, different perspectives have emerged which have formalised diverse meanings of culture ranging from omnibus characterisation of material and non-material accomplishments to meaning or interpretation. These perspectives have institutionalised disparate traditions of inquiry. The disparity is partly prompted by the formal differences across academic interest groups (for example, sociologists, anthropologists, psychologists, experts in cultural studies) but is also rooted in deeper variformities. It seems, that culture is partly a social construction and its meaning varies with the purpose of its user. It has been frequently used to refer to the customs and meanings of particular peoples, variously defined in ethnic, linguistic, religious, political, and socio-historical parameters. Perhaps the definition of culture depends on the proposition or theory about those aspects of the world one is interested in.

Today culture, for various reasons, has become a matter of concern for everybody, including the general public, policy makers, writers, academicians, as well as those who are professionally involved with the theory and practice of culture. This is much more pronounced in the contemporary period than at any other point of time in human history because the science and technology of inquiry is much more formidable now, and the socio-political alignments and configurations are assuming complex and new dimensions. The societal scenario in countries with multiple ethnic groups often involves power play in which certain groups are marginalised. The complexity of culture is reflected in the use of terms such as "high culture", "public culture", "counter culture", "mass culture", "popular culture" etc.

It is interesting to note that human beings are "creatures" as well as creators of culture. Perhaps we are the only species that can establish its existence beyond time and space through cultural artifacts and through local and immediate as well as global and transcendental engagements. We have developed techniques, which while being part of the culture are able to carry, conserve and advance it. In this context culture assumes multiple meanings. It refers to ideas, activities, and objects—all at the same time. It shares and enters all these and is structured through our participation in and through it. As creatures we live in an environment transformed by the artifacts of previous generations. The accumulated system of artifacts co-ordinate human beings with their environment. This gives culture a historical character. Culture may be considered as an attribute of its members or participants. It is a super-trait of humans and the group is its vehicle. As a function of the collective of human brains, the totality of their psychological processes that make up human culture (the shared meanings and practices) is rooted in the process of extragenetic transmission.

Culture is also treated as a process of tending or helping things to grow. It is central to any debate on the nature-nurture controversy. As such it enters as a powerful force of nurture. In the form of preferences and constraints, culture is situated within cognition that, in turn, is present in individual development, which is nested in the evolution of the species. There is incessant and cumulative modification of knowledge and practices over generations. This brings "culture" closer to the notion of tradition and change in modes of thought and style of life.

The importance and magnitude of culture as a phenomenon provides compelling reasons for wanting to understand it in a larger context. Culture is one of the most important substantiations of the maxim. The collective of humans is more than the mere sum. It is, therefore, interesting to ponder over the question of what constitutes culture, what the effects of culture are on individuals and on their world, and what psychological processes make the construction, change and reconstruction of culture possible.

There is a long history of people thinking of culture and psyche being closely interwoven. Attempts have been made to use one to investigate or understand the other. Many have thought of uniform and invariable principles and argue in favour of psychic unity. The cultural differences observed in behaviour are often explained on the basis of climate or ecology. The question of how personality reacts to culture has led to many interesting ideas like basic personality and modal personality. Psychology as a modern scientific endeavour, however, has at times actively dismissed or did not consider culture as an integral part of psychological phenomena. Culture and psychology, for a long-time, remained intimate strangers. The overly scientific stance of the mainstream psychologies did not allow culture to develop as a relevant concern in psychological investigations. It was treated as a source of noise by the psychologists who searched for exact rules or generalisations guiding behaviour. The positivist spirit and the theory of evolution had supplied the basic framework for most of the psychological researches. This tradition is now undergoing a change and other perspectives focusing on interpretation and narrative construction of reality are emerging.

Recent efforts to examine the interface between culture and psychological phenomena have led to the development of many vantage points. In many of these efforts culture has been treated as an antecedent and/or contextual variable proposed to enhance the explanatory power or to account for the variations in behavioural manifestations. Greater concern is shown with the use of culture as a nominal category and a packaged variable. People belonging to a given culture are considered as a homogenous group and are compared with similar group(s) from other culture(s) to decipher the degree of variation in the assumed universal psychological processes. This way of examining culture as a treatment is an extension of the experimental paradigm, and does not offer a comprehensive perspective on the meaning and process of culture and its reciprocal relationship with psychological processes.

Contemporary psychology's concern with culture is voiced mainly by the researchers working in fields like group processes, psychological development, cognition, human resource development and management, stress and health, and the study of self. There is a clear indication of discomfort with the mechanistic, reductionist, individualistic, and mono-cultural practice of Euro-American psychology in the non-western world. The mapping of social reality with the conceptual categories and methods drawn from alien cultures is being criticised on many grounds. The issue of culture-specificity and culture-generality of psychology as a science naturally comes up in this context. However, majority of the researchers and practitioners are at a loss in their attempts at repositioning psychology in the cultural context. The fear of becoming unscientific and losing connections with the mainstream of psychology, coupled with the danger of disintegration of the discipline, has led to a kind of ambivalence among them. As a result, most of them are continuing with the practice of psychology in the usual fashion while a small minority has become interested in reshaping their concerns and activities keeping the cultural context in mind.

In the practice of mainstream psychology culture and social context have often been treated as external entities that function independent of individual action which must be controlled or explained, rather than as a part of the very process of human functioning. Cultural variability has also, in some cases, been quantified using a series of comparative dimensions. However, the implication of the concept of culture as homogeneous and monolithic cannot be sustained since cultural identities cannot be described in terms of fixed categories. More recently globalisation has led to a hybridisation of cultural practices and meanings, which may create multiple identities. Such an enmeshed process of identity construction may reveal that individuals and groups from different cultures could hold similar yet distinct views about, say, a social ritual or an emotion. A variety of approaches for the centrality of culture in understanding the production of human action are required.

While challenging the received view of science, social scientists are engaged in re-examining the purposes and methods of social inquiry. The post-positivist intellectual climate presents an opportunity to think afresh about conceptualizing the culture-psychology interface.

5

Towards an Indigenous Perspective*

The current western thinking about the science of psychology in its prototypical form, despite being local and indigenous, has assumed a global representation and is now treated as universal or the psychological, claiming a pan human level legitimacy. Its dominant voice subscribes to a decontextualized vision with extraordinary emphasis on individualism, mechanism and objectivity. This peculiarly western mode of thinking is fabricated, protected and institutionalized through representational technologies and various scientific rituals. It has been transported on a large scale to the non-western societies under politico-economic domination of western countries. As a result the mainstream psychology maintained a cultural stance at the cost of ignoring other substantive possibilities from disparate cultural traditions. Mapping reality through western constructs has offered a pseudo understanding of the people of alien cultures and had debilitating effects as far as it overshadowed and misconstructed the social reality and demanded unwarranted devaluation and stigmatisation of indigenous concepts and experiences. Consequent to this, the people are facing a dilemma about their identity, their symbolic and conceptual repertoires and behavioural options are being questioned, and a pestoferous academic culture is emerging.

* The ideas presented developed in collaboration with Prof. Ajit K. Mohanty.

Psychology in Modern India

The discipline of psychology as developed in India is primarily based on the knowledge and know-how imported from the Euro-American tradition as a part of a large scale transfer of western knowledge and education (D. Sinha, 1966). Being the recipient it was significantly subordinated to the donor country. The colonial conditioning of Indian mind led to gross neglect and avoidance of psychology in the intellectual and cultural tradition in India which has been shaping the life and social practices of the Indian people since long. The academic world maintained a distance with its own cultural heritage and looked down at it with considerable awe and suspicion. The colonial incursion was so overwhelming that while western concepts were accepted and welcomed without any scrutiny, indigenous concepts were denied entry to the academic discourse. Being imitative, the discipline's growth remained always one step behind the developments taking place in the donor country.

Since its inception as an academic discipline in 1915 when teaching of psychology started at the University of Calcutta, psychoanalytic and experimental psychological streams have received the attention of Indian scholars. Its growth as a discipline and profession was sluggish during the British period. Uncritical acceptance of the positivist and empiricist form of psychological science in the natural science tradition nurtured ritualistic and stereotypical studies of individual (ageless organism) legitimized through quantification. A large majority of such studies illustrates nothing but psychological imperialism and proliferation of an ethnocentric western psychology. After gaining political independence in 1947 noticeable changes started taking place and marked growth in the quantum of research publications can be seen after 1970.

Unlike in the West, psychology in India did not grow as an integral part of the evolutionary process. Being trained by British or American psychologists and with a colonial hangover there has been a strong tendency in the academia to engage in the practice of culture blind psychology. Surprisingly enough this did not create any visible discontentment for a long time. Perhaps the researchers were confident that they were contributing to the cumulatively growing pool of universal knowledge. Thus deviations were treated as errors and problems and issues were filtered through the scientific lens of the West.

For a long period, psychology taught in the Indian universities remained purely western psychology and attempts were made to safeguard it from the possible contaminating effects of Indian culture particularly its religio-philosophical thought. Its teaching maintained a strong universalistic stance. The research studies largely focused on testing the adequacy of western theories and concepts wherein the subjects provided objective behavioural data. In this scheme of scientific activity, culture was an irrelevant and extraneous factor in the genuine search of behavioural laws. The S-O-R framework till now has furnished the basic paradigm for psychological research. Such a practice of psychology was non-reflexive and therefore, could not recognize the discontinuity between the native knowledge system in India and the (western) psychological conceptual categories. The two seemed to belong to unrelated genre, one being mere speculation while the other grounded in hard core objective facts.

It was only during 1970s that Indian psychologists started taking note of the problems in terms of methodological artifacts and incompatibility between theories and Indian reality (Mukherjee, 1980; Rao, 1986; J.B.P. Sinha, 1984). During 1980s signs of disillusionment with the state of affairs emerged (Mohanty, 1988) and innovative approach and indigenous thinking became visible. Attempts to go beyond emulation of western models also started (J.B.P. Sinha & Prakash, 1993) and value of indigenization was gradually recognized (Adair, Puhan & Vohra, 1993; Adair *et al.* 1995; D. Sinha, 1994). In recent years scholars have started moving towards reorienting psychological discourse by locating it in the culturally rooted conceptual space. This has been facilitated by interactions of Indian psychologists with cross-cultural psychologists.

Culture and Psychological Processes

It is being realized that psychological processes are rooted in historically variable and culturally meditated practical activities. People construct multiple socio-historically grounded realities and apprehending reality from an observer-independent perspective is virtually impossible. Cultural psychological studies underscore the constitutive role of the context of activities. We as human beings operate within socially constituted worlds. This defies the assumptions of a lawful universe and absolute objects with context-free properties. Investigations from a culturally rooted perspective tend to show that many western concepts lack experiential validity in other cultures. They offer evidence that the notion of psychic unity leading to presumed universal and identical psychological phenomena or processes (e.g. self, emotion, morality, well-being, development) is not viable. Instead, they are constituted differently by the varying socio-cultural contexts. Here special mention may be made of the modes of self-construal which provide an interpretative framework shaping the way a given socio-cultural group engages in thinking, feeling and acting. Markus and Kitayama (1991) have marshalled evidence that interdependent and independent modes of self-construal have diverse implications for defining and shaping psychological processes. Such construals simultaneously constrain and afford a range of psychological processes which members of a given culture may have developed through socialization in order to conduct themselves. Thus being unique as well as interdependent both became equally valuable cultural imperatives in respective cultures. Being independent sets an agenda for a person quite different from the tasks ahead for a person holding an interdependent self-construal. Diverse self-construals encourage different sets of customs, practices, artifacts, norms and institutions in the society and are also sustained by them. Some of them remain explicit while others remain less conspicuous but either way they constrain and afford differing options for human action. It is in this sense that psyche and culture jointly make each other up (Shweder, 1990).

Interestingly the modernist view of individual as a self-determining and autonomous being is losing its functional value as the post-modern conditions necessitate a global functional unit. In this context as Sampson (1988) argues the theory of person needs to be revised. To this end he endorses a constitutive view which holds that persons are viewed as guardians of particular assets of the community, and not their owners. Concomitantly there is a resurgence of interest in approaching human phenomena through a human mode of thought, and issues of subjectivity, interpretation, and everyday understanding are becoming salient. The cultural embeddedness of thought and action is being emphasized. This change signifies the possibilities of developing a situated and shared conception of reality. It goes beyond the positivist position and proposes that the knowledge claims in the human domain are relative to the setting in which they are grounded.

The Move Towards Indigenous Psychology

At a metatheoretical level, Paranjpe (1984, 1998) has shown the possibility of relating eastern and western conceptualisations of self, identity and consciousness. Misra and Gergen (1993) have explored the possibility of articulating Indian (Hindu) construal of psychological functioning. They have noted that the ontology of personhood in the Indian cultural context is rooted in both the spiritual as well as the natural worlds. Looking from this position and informed by the available analyses, the following features emerge: holistic-organic world-view: coherence and order across all life forms, socially constituted/embedded nature of person; non-linear growth and continuity in life; behaviour as transaction; temporal and atemporal existence of human beings; contextualized action depending upon time, place and person; search for eternity in life; emphasis on self-discipline; transitory nature of human and physical experiences; distributed, rather than personalized control; belief in multiple worlds (material & spiritual) and action as the locus of cause and effect.

Granting the above description as an approximate portrayal of a person and his placement in the social and physical environment, implies that person and social situation both mutually define each other. It is also evident that instead of searching for a simple cause-effect relationship, a context dependent or inclusive strategy would be more desirable. From the standpoint of such a view of a person, the role of psychologists would be to help in understanding, reading and interpreting the behavioural episodes and informing about the potentialities of action within the range of possibilities available in the culture. In this scheme, the researcher will be a co-participant in the joint construction of future course of a person's behaviour. The discursive mode of inquiry would be more suitable in the Indian socio-cultural context in which text and context exist side-by-side and influence each other. As students of culturally constituted psychological phenomena, we think that the tool-box or research instruments as well as the dimensions of discourse within the discipline warrant innovative reconstructions. In particular, different forms of linguistic products and activities would require greater attention, not as representations of some underlying mental phenomena, but as real human activities. This will expand not only the repertoire of our analytical tools but also enrich them by adding new dimensions to the theoretical and conceptual arena of the discipline. Such an effort, of course, would be cross-disciplinary and utilize the resources from allied disciplines. This kind of participatory practice would be creative and emancipatory and would go a long way in enriching the indigenous cultural traditions.

There is now a discernible shift in the academic orientation of Indian psychologists in which cultural sensitivity and social relevance have become prominent (Pandey, 1988; Paranjpe & Ribber, 1988; Rao & Sinha, 1997; Misra & Varma, 1995). This has resulted in questioning of western psychological constructs and methods of explicating and understanding the Indian reality. These efforts to offer alternative construals of social reality have taken various forms including theoretical and methodological innovations in social, psychological, clinical and organizational contexts (Misra, Prakash & Varma, 1999; Varma, 1995). A fruitful interface between the indigenous Indian thought and psychological discourse is found in many recent moves in psychological applications. *Guru Chela* paradigm of therapy (Neki, 1973), nurturant task style of leadership (J.B.P. Sinha, 1980), analyses of self and personality (Nandy, 1983; Paranjpe, 1984; Rao, 1988b; Sinha & Naidu, 1994; Tripathi 1988); conceptualization of achievement (Misra &

Agarwal, 1985; Dalal, Singh & Misra, 1988), analysis of Indian psyche (Kakar; 1995, 1996), emotion (Jain, 1994), justice (Krishnan, 1997), morality (Misra, 1991), concept of well-being (Sinha, 1990a), development (Saraswathi & Kaur 1993; Saraswathi, 1999), values (Prakash, 1994), detachment (Pande & Naidu, 1992), and methods of organizational intervention (Chakraborty, 1993) are some of the important examples. Also, recent research has helped in providing analysis of moral order (Shweder, Mohapatra & Miller, 1987; Miller, Bersoff & Harwood, 1990), self (Roland, 1988; Marriott, 1976; Misra, 2001; Misra & Giri, 1995) and socialization (Kurtz, 1992; Seymour, 1988).

As Marriott (1992) has pointed out "alternative social sciences are potentially available in the materials of many non-Western cultures, and their development is essential to serve in the many places now either left to *ad hoc* descriptions or badly monopolized by social sciences borrowed from the West" (p. 269). Disciplinary socialization like any other kind of socialization has deep-rooted effects and its products and byproducts tend to maintain a given sub-culture. Indian psychologists are facing a similar problem. Being decentred, the Indian perspective goes beyond the scientific view and is therefore rendered 'philosophical'. For this reason, it is treated as unsusceptible to the observational language of western psychology and is therefore considered a non-entity.

The rejection of western models and lack of confidence in Indian cultural thought contributes to an existential threat. The solution of this problem can be found in respecting both the traditions. About four decades back, Sinha (1965) had indicated a need for integration of modern psychology with Indian thought. Since then a variety of efforts have started. They have yielded some interesting results ranging from revitalization of traditional Indian concepts through use of a mix of western and Indian concepts. In general, a relational constitutive view of a person is emerging in these studies. All this is largely done within the western scientific paradigm. Hopefully, Indian psychology will outgrow this transitional phase and overcome the pains of alienation. This, however, would require "a thorough relativization of the western concept of psychological science and indigenously based formulations of both individual functioning and of scientific inquiry nurtured into full maturity" (Misra & Gergen, 1993 b.p. 254).

In the above context, indigenization becomes a process of making knowledge culturally appropriate. A truly indigenous approach, therefore, makes psychology embedded in specific cultural contexts and facilitates establishment of universality of its principles. Such a move is now quite widespread across the globe, particularly among the decolonized countries. For India indigenization means integration of modern psychology with Indian thought. Western psychology has been ethnocentric in its practice. The growing disillusionment of foreignness of psychology and realizing the conceptual and methodological constraints in its application has been the major impetus for indigenization.

Indigenous Psychological Orientation

The development of indigenous psychological orientation in psychology is a reaction to the unjustified claims of universality. In this approach culture is not viewed as a variable. It is considered as an emergent property of individuals interacting with their environments. As a superordinate concept culture offers a framework for viewing what is relevant and meaningful. In indigenous approach to psychology concepts are locally defined and provide a contextually meaningful and problem oriented perspective. As Sinha (1997, p. 132) has indicated the main features of this emerging approach include the following:

(1) Psychological knowledge is not to be externally imposed; rather the cultural tradition should give rise to it.
(2) True psychology lies not in artificially (experimentally contrived) induced behaviour, but in daily, mundane activities of the people.
(3) Behaviour is to be understood and interpreted not in terms of imported categories and foreign theories (i.e. imposed etics), but in terms of indigenous and local frames of reference and culturally derived categories.

(4) It embodies psychological knowledge that is relevant and is designed for the people. In other words it reflects the socio-cultural reality of its society.

In recent years the issue of indigenization has become an important theme in the discources related to critique and reconstruction amongst psychologists in different parts of the world. The disillusionment with the theories and practices of western psychology has resulted in the development of indigenous psychological approaches in Africa (Durojaiye, 1993; Nsamenang, 1992; Ogendenngbe 1993), China (Bond, 1986; Ma, 1988; Yang, 1982; Wang, 1980), Japan (Dol, 1986; Hoshino & Umemoto, 1987; Nakane, 1970; Kitayama & Markus, 1999; Yamaguchi, 1994), Latin America (Ardila, 1993; Diaz-Guerrero, 1982, 1993; Diaz-Loving, 1994) and Philippines (Enriquez, 1982, 1992). In these and other similar efforts (see Kim & Berry, 1993; Sinha, 1997a, Rao & Sinha, 1997) attempts are in progress to develop indigenous approaches in various domains of psychology. On the whole it is still in the stage of infancy and involves diverse perspectives and modalities reflecting different types and levels of indigenization (see Sinha, 1997b). In a recent publication Kim, Park and Park (2000) have tried to articulate indigenous psychology as a "transactional paradigm in which context, epistemology and phenomenology are central aspects of research" (p. 71). Its core assumptions are;

(1) Subjective and objective perspectives are not separate realities, but they represent two interrelated ways of understanding reality.

(2) Research is recognized as a value-laden enterprise in search of a probabilistic understanding of the world rather than a search for objective, deterministic, and mechanical knowledge.

(3) Research topics and stimuli must be meaningful and contextualized rather than created arbitrarily by researchers.

(4) Rather than instinctive and reactive, people are recognized as interactive and proactive agents of their own actions.

The meanings of and reactions to the notion of indigenization of psychology has also drawn attention. While there are no well formulated ways and means for indigenization this movement has drawn the attention of psychologists interested in making psychology culturally relevant in many directions. Its practice has many variations and the issues have been approached from different perspectives leading to wide ranging connotations. There is no single route towards achieving this goal. It is a multi-dimensional and pluralistic exercise. On the one hand, it involves attention to the traditions and roots of Indian culture and on the other it has to respond to the contemporary realities. Thus a kind of blend of integration of Indian thought and western knowledge is desired.

Some researchers focus on the indigenous conceptual resources and present efforts by psychologists in the direction of indigenization. The researchers draw inspiration from various sources and venture at generating more efficacious, comprehensive and at time fresh perspectives on a variety of topics (e.g. knowing, social cognition, emotion, stress, well - being, the notion of self and personality). They are influenced by the richness and depth of Indian thought systems and their potential to facilitate broad based solutions of human problems. They offer a new contextualized vision on reality and understanding.

Psychological issues in cultural context confront the challenges in overcoming a western ethnocentric psychological science. In so doing they point out the centrality of culture in pursuing any culturally relevant psychological inquiry. The symbolic resources and practices of various cultures construct world-views and enable us with particular viewpoints to appreciate reality and manage our conduct. In today's world where diverse communities are trying to assert their social identities it is crucial that we approach cultural diversity in the sense of an opportunity and try to respond to this challenge in a constructive way. Indigenous psychology will go a long way in furthering such an approach.

In contemporary disciplinary discourse regard for cultural variety is gradually gaining momentum but the vision of psychology and its mechanisms are still largely informed by the western natural science model which redefines the psychological processes which are essentially discursive in nature. The cross-cultural psychology is suffering from a tension of maintaining its western stance and hopes to incorporate diverse cultural realities in its fold. In this context many biases still prevail. For instance Triandis (1993) claims that 'western knowledge system is a precondition of doing science and traditional India had no science'. This reflects ethnocentrism and lack of familiarity with other cultures. The history of science shows that pluralism and diversity have been healthy for its growth. The unsolicited advise of Triandis that 'if there is to be an Indian science it has to follow Western patterns' is something which Indians have become aware of on account of colonial experience and western educational training. The result has been the development of a psychology which is 'completely isolated from the Indian tradition and alien to the local intellectual soil' (Sinha, 1986, p. 11). The western empirical-positivist mode is not suitable to encompass the entire range of psychological reality because it does not mirror any absolute reality and is insufficient to tackle human reality in which meaning mediates actions which in turn is rooted in cultural practices. Culture antedates behaviour and it is through culture that a behaviour is constituted and articulated in some sensible fashion. We need to see the continuous and dynamic interface between culture and psyche. Since cultures are not static and closed we need to attend to historical and social structural processes also.

6

Understanding Human Development*

The scientific study of human development, beginning almost a century ago, has undergone significant transformation. At present this field is characterized by a multiplicity of theories and methods focused at uncovering the nature, and mechanisms of developmental change in different realms of human functioning in the life course. Historically, however, 'development' was viewed as an orderly unfolding in a predetermined sequence, with possible individual differences in the timing of manifestation of various universal characteristics. The prefigured ideal outcome or endpoint postulated in the preformation and epigenetic views has been challenged by the effects of environmental interventions documented in various studies.

Under the impact of natural sciences and zeitgeist of reductionism, students of living systems adopted a materialistic view, which conceived developmental complexity in terms of accumulation of simpler behaviours at lower age levels. The emphasis was given to quantitative development. In this endeavour, the qualitative aspects of development were also reduced to quantitative changes. For instance, mental growth was treated as outcome of learning which can be observed and controlled (Sears, 1957). An alternative approach was also proposed which considered development as age-correlated change. This was subsequently refined by emphasizing age-related changes in various domains of

* The ideas presented developed in collaboration with Dr. K.N. Tripathi.

development. Following this perspective, Wohlwill (1973) identified the following tasks for the study of human development; (a) parametric description of age-correlated change, (b) specification of antecedent-consequent relationship at a causal level, (c) validation of process-oriented models using multivariate systems, (d) discovery and synthesis of developmental changes, (e) descriptive study of age-characteristics, (f) correlational analysis and (g) study of determinants of development. These goals have been guiding the work of researchers in the field of human development, and age-related changes in physical, psychological and behavioural domains have been frequently mapped.

The theoretical models of human development, proposed earlier, may be broadly categorized as mechanistic and organismic. The mechanistic model viewed organism as a machine and considered it reactive and empty. It followed an epistemology of naïve realism, and emphasized end-products and quantitative changes. In contrast, the organismic model conceptualized humans as living systems in which parts are related to the whole. It had an epistemology of constructive realism and delineated the qualitative changes. Both these perspectives have led to studies of development with a focus on time-ordered changes or changes that occur with age. More recently, focused attention on the role of environment in explaining development has led to the emergence of a contextualist perspective.

Environment as the Context of Development

The use of environment in developmental theorization has been attempted in many ways and with different implications. Following Wohlwill and Heft (1987), four major perspectives are identifiable in this regard.

Stimulation—For a long time environment-development relationship had presumed independent impact of the social and the physical environments. Hebb's (1949) insights into organismic functioning led the attention of researchers to deprivation-enrichment dimension of environmental stimulation. Wachs (1977) proposed that we need a concept of optimal level of stimulation, dependent on the organism's level of development (that is, prior history and past achievement). Environmental stimulation is thus viewed as reciprocal and mediated by the actions of the organism. The environmental variables affect development through a filter or mediator (e.g., cognition, adaptation level). Here, it may be noted that, since physical and social environments are embedded in each other, an interactional model appears to be more viable. Recently, transactional models have been proposed which consider the mutual definition of person and environment, and emphasize the non-linear causality. This view considers the environment-behaviour, as a single unit of analysis. The transformational model advanced by Stokols (1988) is an extension of the transactional model, which puts emphasis on the dynamic processes of change and transformation as a fundamental aspect of the environment-behaviour development relationship. In addition, it has the following features: social and physical aspects of behaviour settings are highly interdependent, occupants are actively involved, and people-environment relationships are reciprocal and dynamic. Magnusson (1988) has used the interactional perspective in a study of individual development longitudinally over a period spanning two decades. This study showed that individuals develop and function as integrated systems. There are continuous reciprocal processes of interaction among subsystems of psychological and biological factors. The function of the total systems is often non-linear.

Context—Bronfenbrenner (1979) has viewed environment in terms of a nested hierarchy of micro, meso, exo and macro systemic contexts. These different subsystems influence each other in an interactive fashion. Stokols (1983) has proposed an approach which includes spatial, temporal, and sociocultural contexts of the phenomena being studied. In this approach the key physical and social environmental variables influence development in a given situation, independently as well as interactively, together with additional affective contextual variables.

Source of feedback—Some researchers emphasize the fact that the correlation between motor activity and its consequence provides feedback to the organism. Thus, environment serves as a source of feedback. This feedback is often mediated by the social and cultural context. The participation of children in different activities is usually guided by parents and elders in the family that help internalizing the tools for thinking, cognitive activities and related skills. This view is reflected in the concept of *'zone of proximal development'*. As Vygotsky (1978) has argued, the zone of proximal development is very crucial in which development proceeds through children's participation in activities with more experienced members of their culture. Thus children and their parents develop, from their interactions with each other, a derived understanding based on their efforts to apply the tools of culture. Rogoff (1990) has recently extended the concept of the zone of proximal development by emphasizing the interconnectedness of the roles of children, their care givers and other companions and the importance of diverse types of face-to-face social interactions in guided participation in ongoing cultural activities. Thus children develop while participating with other people in a social order. This position emphasizes shared problem solving as a major factor promoting cognitive development.

Set of affordances—Gibson (1982) has argued that environment furnishes various opportunities and resources which are functionally relevant. The meaning of functional possibilities of environmental features is therefore very crucial. The affordances have relational character and, therefore, have a developmental dimension. As Heft (1988) has shown, the functional taxonomy of environment is psychologically more meaningful than mere description and enumeration of environmental stimuli.

Environment-development relationship is now viewed in a contextual perspective, which takes the stand that a continuous and necessary transaction takes place between the organism and its environment (Sameroff, 1983). This view is further enriched by cross-cultural investigations, as well as sociological and anthropological studies of modernization and change. It has become clear that age-related developmental changes cannot be fully understood unless one takes into account the social time and historical time. The chronological age in itself does not tell much. It is a non-causal variable and works as a marker along which other experiences occur. More specific underlying processes need to be explicated for understanding developmental changes. Social time is defined in terms of patterns of events an individual passes through while encountering a variety of social institutions in a given culture. The historical period, characterized by socio-political, cultural and economic happenings, exerts great impact on development. The developmental stages and participating institutions take different shapes during different historical periods. Thus, development needs to be understood in its context which also changes with time.

Patterns of Developmental Change

Past research has documented that developmental changes take different forms. They may be continuous or discontinuous, depending upon the period in life course. For instance, childhood usually shows incremental change while adulthood is characterized by irreversible decrement. The traditional developmental theory, however, could not adequately explain developmental changes during adulthood and aging, as it was preoccupied with *stability* and *ordered* (sequential) *development* premises. Neither longitudinal studies nor life histories could justify the stability account. On the other hand, the orderly change account found in the works of Piaget, Erikson, and Freud is also not consonant with data which often show reversals and regressions. Adult development, as characterized by Gergen (1982), is *aleatory* (chance-dependent), and there is little about human development that is pre-programmed or prefixed. The human life course is chiefly open-ended, and stability and orderliness are largely limited to the early stages of development.

Human beings have great capacity for change during the entire lifespan. The patterns of change across lifespan, therefore, are non-recurring and multidirectional. The traditional approaches to human development could not take into account this complexity. Also, the emerging model of human beings which favours autonomous, self-organizing and active organism was not congruent with the earlier approaches. As Reese and Overton (1970) stated, human beings are sources of acts rather than collections of acts initiated by external forces. They, as active participants, are characterized by internally originating change. The limitations of traditional developmental approaches were overcome by extending the strategy of study to encompass the entire lifespan.

The Lifespan Perspective

As stated earlier, the predominant strategy of development studies was to look for developmental changes in a contextual framework. As such, it was least sensitive to the role of historical experience of the developmental system. Only recently, developmental researchers have started investigating age-related changes as they occur over the entire course of a person's life, and to situate these changes in a cultural-historical frame of reference. Such an approach emphasizes the interconnectedness of age periods and recognizes reciprocal embeddedness of a person in a social context, by postulating that development is culture-bound, and constituted and regulated through individual and social action.

The focus of this approach is on the study of development within multiple contexts including the life course, the social context, and the cultural and historical context. It may be noted that these contexts represent dynamic processes which have a developmental history of their own. Also, individuals and social systems both simultaneously and continually grow and develop. Thus by expanding the scope of development, this perspective questions the presumed locus of development within the individual. To this end, it directs attention towards the socioecological matrix within which human development takes place and looks for several interrelated foci of influence. Idealy, this approach strives for a comprehensive mapping of the dynamics of development across age, life events, and history, and exemplifies joint application of idiographic and nomothetic approaches. According to Hetherington and Baltes (1988), it is based on the following propositions, which distinguish it from the other perspectives:

- Ontogenetic development is a lifelong process and no one period in life is primary. Thus joint occurrence of gains (progression) and losses is involved in the course of development.

- The direction of change varies across categories of behaviour. Thus it is multi-directional.
- Much intra-individual plasticity is found in psychological development. Coordination across areas, persistence, and interdependence in development, therefore, become the major issues for study.
- Ontogenetic development can also evince variations in accordance with historical and cultural conditions.
- The pattern of development is characterized by interaction of age-graded and non-normative factors.
- Psychological development should be studied in the interdisciplinary context informed by other disciplines concerned with human development.

The lifespan perspective has been taken up by many researchers studying different aspects of development. For instance, Lerner has examined personality development in a lifespan perspective. He argues that an adequate conception of personality involves universal, subgroup, and idiographic components of the person. Separately and in combination, each of these components assures the lawful individuality of the person, his or her reciprocal embeddedness in a social context across life and, as well, his or her relative plasticity and openness to this context across life (Lerner, 1988, pp. 41-42).

This kind of approach to personality development encourages longitudinal, change sensitive, multilevel and contextual analysis. Another growing concern of lifespan researchers is to include family in its analysis. Families are composed of sets of interlocking relationships. This expansion necessitates inclusion of different levels of analysis. Parke (1988) suggests that dyadic (e.g., parent-child, husband-wife and sibling relationships), triadic (e.g., parent-child-sib, mother-father-child) and familial levels of analysis be added to the developmental scenario in order to trace the ways in which these differing units change across developmental stages and interact to produce diverse effects. Historical context is also a necessary ingredient in this analysis. Family development deals with the process of progressive structural differentiation (Hooper and Hopper, 1985).

Some of the important contributions of the lifespan perspective are found in the study of intellectual development. As Sternberg (1988) has pointed out, such studies indicate that intellectual development does not end in adolescence, and adolescence cannot be viewed as the pinnacle of intelligence. Also, informal knowledge is important for intelligence. The traditional evaluative criteria for theories and tests of intelligence prove too narrow to encompass the changes during lifespan. Lifespan studies clearly indicate that intelligence resides in the interaction between the individual and the environment. Finally, the cohort effects apply to children as well as adults. On the whole, it appears that the lifespan perspective is more optimistic about the malleability of human beings and has greater scope for modifications of the effects of early experience. Also, the large scale heterogeneity during any age level suggests that age linked criteria would not suffice. This becomes crucial as the accumulating evidence suggests that different behavioural domains have different rates and features of development.

Thus the lifespan perspective has many advantages over the traditional perspectives on development. However, its practice is quite challenging and many issues have to be sorted out in order to use it fruitfully. For want of space, only one important issue is being singled out for attention. This issue is of integrating different disciplinary perspectives and levels of analysis. As one can easily discern, the lifespan perspective does not have a neatly demarcated area. Its research issues often cannot be adequately analyzed from single theoretical and/or disciplinary orientation. The specialization and theoretical orientation(s) prevailing in various disciplines (e.g., psychology, sociology, anthropology) lead to diverse perceptions and constructions of research problems. The level of analysis particularly manifested in the unit of analysis (e.g., response, individual, group, subculture, culture) chosen by the investigator delimits the range of empirical referents and the patterns of possible causal analysis. It may be noted that usually one level provides a context for the other levels and each behaves as an independent variable in relation to the others. These levels also involve different time dimensions. The integration of perspectives and units of analysis is complex.

However, single level analysis always has a greater chance of leading to trivial, tautological or arbitrary propositions (Allardt, 1969). In contrast, the generalizations containing variables from different levels will be more specific.

The problem of integration has been addressed in several ways including *aggregate analysis, ethogenic analysis* (Harre, 1979) and *multiple triangulation* (Denzin, 1970). Aggregate analysis stresses the integration of levels. It subscribes to the view that inferences can be made about one level of analysis from data collected and measured at a different level. The ethogenic approach emphasizes the integration of concepts and disciplinary and theoretical perspectives relevant to a research problem. Multiple triangulation occurs when researchers combine in one investigation multiple observers, theoretical perspectives, sources of data and methodologies. This strategy captures a more complete, holistic and contextual portrayal of the unit(s) of study. Thus it has greater scope for convergent validation, uncovering deviant cases, and generation of hypothesis. The specific mechanisms for integrating different levels of analysis and disciplinary perspectives in general, and human development in particular, are yet to be worked out.

Human Development Research in India: An Appraisal

The study of human development in India is a fascinating and challenging task. The rich textual heritage, cultural and subcultural diversities, geographical variations, colonial background, and changes resulting from vigorous efforts towards modernization, industrialization and urbanization have made the tasks of specifying Indian ecology of human development very complex. In the last few decades increasingly greater attention has been paid to the issues pertaining to human development by Indian psychologists, pediatricians, sociologists, anthropologists, and students of other disciplines, and sizable data have been generated which throw light on different aspects of development. The four surveys of the field (Anandalakshmy, 1980a; Kaur and Parmeswaran, 1972; Saraswathi and Dutta, 1988a) indicate that there is clear growth in the quantity of studies and a change in research interests in favour of investigating problems of applied significance. Notwithstanding this positive sign, it is also noteworthy that all is not well with the discipline (see Saraswathi and Dutta, 1987, for a critique).

A critical look at the scenario of human development research in India yields a mixed picture of certain strengths and many weaknesses. The positive aspect, although representing a microscopic minority of researchers, includes increased sensitivity to contextual variables, a concern for societal problems, certain degree of openness to inputs from other disciplines, methodological innovations, and interest in identifying indigenous concepts. This represents a new research orientation and evinces a small but significant step in the direction of promoting a relevant and sustainable discipline of human development. The central tendencies and modal trends, however, are not so encouraging. Lack of professionalism and limitations of the prevailing academic culture have given rise to the emergence of several practices that are inimical to the growth of the discipline.

The theoretical foundations of the studies are usually not explicit. However, an analysis of interpretations advanced by the researchers gives the impression that the maturational view characterizing age-correlated normative progression continues to form the predominant framework. It is only recently that the context of development is being recognized for explaining developmental changes. The implicit model of the human being, adopted in these studies has been that of self-contained individualism. Human beings are viewed as manipulatable objects. Unfortunately, the self-reflexive critique and debate at the conceptual level has been largely absent. One outcome of this state of affairs has been the uncritical use of alien concepts to describe and explain various phenomena as if they were universals (or etics). In actuality they were imposed or pseudo-etics, and therefore, could not provide a genuine description of reality. Most of the variables relevant to development are rooted in specific cultures (e.g., child rearing, family structure, roles and relationships, values and norms). Very little effort has been made to locate and identify the culturally specific (emic) features of the Indian context of human development.

The general methodological preference in Indian studies

has been to use chronological age (and sometimes grade) as a treatment or an independent variable, which leads to findings such as 'children of different age groups are different because of differences in their age'. Age is used as a packaged variable and no effort is made to unpackage it. As a result, one never knows the mechanisms/processes which have led to the changes observed in the children of different age groups. Another methodological practice which needs mention is that of instrumental rigidity. Since developmental studies by nature are open-ended, the measures should have scope for allowing variation and a certain degree of openness to encompass new patterns and phenomena. Indian researchers heavily emphasize the prefixed coding categories and are often reluctant to welcome new patterns of behaviours or responses. This is amply reflected in the rare use of the naturalistic perspective in research. The result of this tendency is the maintenance of *status quo* in knowledge and understanding. Also, it detracts one from confronting the reality and contributing to the solution of problems, as the forces favouring *status quo* inhibit any endeavour towards conscientization and/or annunciation (Freire, 1975).

Future Directions

A constructive intervention in the teaching-learning and research strategies prevalent in the discipline is necessary in order to provide clearer directions for the future. Some possible alternatives are delineated below.

Development of a Research Network

One of the major steps that needs to be taken is building a strong research network, so that data may be integrated and made available to the different categories of users. The research questions and priorities have to be spelt out and periodically scrutinized. This would enable useful and productive interface between theory and application, and hopefully replace the lack of systematic problem-focussed research. In order to reorient the teaching-learning process, preparation of textbooks on human development emphasizing the Indian socio-cultural context is a prerequisite. Unfortunately, this task has not received the attention it deserves.

Emphasis on an Indian Perspective

In recent years Indian perspective is being stressed by social scientists in a number of ways and with varying connotations. Here, the Indian perspective should represent the indigenous view, that is, a view with an emic orientation. This needs encouragement to combat the impact of Euro-American theories, which have led to the uncritical acceptance of imposed constructs and realities. So far, the modal Indian response to the Western impact on theory has been of uncritical acceptance and imitation. There have been, however, some attempts to bring into focus the Indian perspective. An interesting example is the work of Misri (1986). Based on fieldwork on Kashmiri Pandits of Umanagari, Misri, an anthropologist, argued that the conception of the child can be reconstructed on three axes which represent antithetical ideas. First, the child is seen to be simultaneously a creation of the parents and a gift from god, the axis of human-divine. It assumes the parents to be mere instruments through whom the divine will is realized. Second, the child is both a collective being and a unique individual. The concept of Karma emphasizes the individual, while relationships with the family and others stress collectivity. Third, the child is not perceived as a blank slate. Though the child is born with fixed and unalterable propensities, his initial nature can be transformed by Sanskaras and rituals. This is the alterable-transformative axis. Within this paradigm, which rests on contradiction, society attempts to simultaneously integrate and separate the new born (p. 118).

Methodological Pluralism

Understanding development across lifespan in its context requires a multipronged methodological approach, which is flexible enough to accommodate participants from different backgrounds and the data having different degrees of structure. Some of the ways to address the methodological challenges are as follows:

- Multiple measures should be used to assess a given phenomenon.
- Greater emphasis should be on the qualitative aspects of the data and cross validation of research findings.
- The independent variables need to go beyond nominal characterization. Terms like 'caste', 'rural', 'urban' and 'tribal' in themselves are meaningless and may not be treated as self-explanatory. They should be clearly specified in terms of some proximal variables. This is also necessary for developing intervention strategies.
- Programmatic research and follow-up studies have to be undertaken.
- The studies should take into account the historical-social context of human development. This is essential for appropriate conceptualization of the individuals who constitute the subjects of research.

Move Towards Non-positivistic Perspective

It is important to note that the present methodology being dependent on positivistic philosophy has led to a pre-occupation with observing the immediate or here-and-now behaviour. As a result, developmental psychologists the world over have become chroniclers of the present. This framework provides verifiable data at the cost of generality. At present in the West there is a growing awareness that, in order to seek the general and lawful, we have to go beyond the positivistic philosophy of science, which involves a relatively fixed and static conceptualization of science. It does not pay due attention to the growth and development of science itself. In other words, it is non-developmental and prescriptive. Recent advances in post-positivistic philosophy are more developmental in nature and reflect more accurately how scientists actually practice science. Historicism and contexualism are considered more critical to this perspective. Also, it emphasizes goal-directed and theory-guided practice (praxis). These trends are more congruent with developmental explanations which are autonomous (non-reducible) and non-causal.

In contrast to the deductive-nomological explanation, many new models of explanation are being proposed (Kitchener, 1983). The theoretical foundations of contemporary psychology are also being critically scrutinized. Search not only for alternative theoretical perspectives but also for expanded epistemological foundations is being vigorously undertaken (see Fiske and Schweder, 1986; Gergen, 1991). A general discomfort over the empiricist meta-theoretical programme is visible, and concern over objectivity and rationality of social sciences, narrow scope of generalizations, lack of paradigmatic convergence, inability to specify the conditions under which events happen and respond to the practical demands of society, and inadequacy of communication of the findings is being expressed. The post-empiricist shift presents several possibilities in which realism and constructionism, with emphasis on cultural contexts and self-reflexive analysis, are most prominent. The related debate makes it amply clear that preoccupation with data collection cannot answer the questions pertaining to developmental analysis and uncovering invariant sequences and transitions (Siegel, *et al.*, 1983). Indian researchers have yet to recognize and respond to these theoretical-methodological challenges.

Sensitivity to Social Change

While research in the field of human development in India has definitely broadened in scope, and investigators have started realizing the importance of orienting themselves to the needs of society, the research priorities, at least as reflected in publications, suggest that there is scope for further improvement. The Indian context of large scale and complex socioeconomic change implicates changes at family, community and cultural levels, which are transforming the ecology of human development in significant ways. As identified by Sinha (1988), these changes include (a) nucleation or the transition from an extended-kin to a primary-kin system; (b) segregation of children from adults, and individuation; (c) change from indulgent to strict child rearing; (d) inconsistencies in child rearing; (e) absence of clear-cut role models; (f) changes in the status, role, and employment of women; and (g) impact of modernization. These points succinctly summarize the changing family scenario in India and raise several issues that command research attention. The process of rapid social change creates pathologies (e.g., divorce, drug addiction, diseases, conflicts and constraints) and requires restructuring of social networks to accommodate the changes (role definition, time allocation, role relationships). Problems of adjustment during adulthood (e.g., unemployment, marriage, friendship, obligations) and old age are also assuming new dimensions. These and other practical problems have many sub-cultural variations (e.g., rural, urban, tribal) which make them interesting as well as complex.

The study of human development has certainly moved from preformationism to contextual and interactive analysis in which non-normative and historical factors are taken into account. Viewing developmental changes in the lifespan perspective replaces stability and ordered progression by multi-directional and dynamic changes in the socio-ecological matrix. This perspective faces the challenge of integrating data of diverse types. The Indian studies, however, have been predominantly concerned with age-related changes, particularly during childhood, and are largely atheoretical and acontextual. They also have methodological limitations. Recent years have witnessed increased awareness of practical problems, sensitivity to ecological conditions and a concern for theoretical-methodological innovations. There is need for developing research networks, emphasizing an Indian perspective, respecting methodological pluralism, and increasing sensitivity to contemporary social change. Human development occurs in an environmental context which is changing and in turn redefining development at the societal level.

7

Rethinking Quality of Life

Search for an ideal life has been a consistent concern through the ages. Efforts towards its conceptualization have been intimately linked with assumptions about the essential nature of human beings and their existential concerns. The expression Quality of Life (QL) has emerged to address empirically the problems of identifying the indicators of well-being of people so that policies for constructive intervention at the societal level can be formulated. QL refers to certain objective societal conditions (e.g., availability of and access to physical, social, educational and economic resources) which are deemed necessary for optimal functioning in a given eco-cultural context. At an individual level, QL refers more to the perception of short-term, long-term or very long-term judgements of happiness and satisfaction which one experiences or expects to experience. As such, it becomes a purely personal and subjective affair. Of course, in both the conceptualizations, the socioeconomic and ideological context to which a person belongs and within which a given society operates, play a decisive role.

A close perusal of the analyses of QL prevailing in contemporary discourse reveals that individualized or personalized need-satisfaction is their central concern. Such depictions portray the celebration of a Narcissistic self. This view ignores certain important aspects of the social reality in Indian and many other non-western cultures. In particular, the constitutive function of our social or community life is often left out. The need to have a more comprehensive notion of QL that recognizes the limits of a person-centred view and proposes a more inclusive and transpersonal perspective, is being voiced.

The Contemporary World Scenario

Considered at a global level, the life conditions in the contemporary world present a dissatisfying state of affairs. Towards the end of the twentieth century, the human condition does not demonstrate any clear improvement over the past. The idea of progress and the dream of unlimited growth have not been successful in improving QL for a large section of the world's population. In the affluent and so-called developed countries, large scale concentration of politico-economic power and physical resources has been able to provide freedom, security, and wellbeing for their people to a considerable degree. Living conditions in these countries apparently present a picture of greater material wealth, high rate of consumption and increase in the duration of leisure time. However, it also exhibits many symptoms of social pathology and psychological disorder. The situation in these affluent nations does not validate a linear relationship between material comfort and wellbeing. The threat to wellbeing is also faced in the third world countries but that is due to abject poverty and social inequality coupled with the ill-effects of colonial past. Most of these countries are now engaged in large scale and rapid process of socioeconomic change through industrialization, urbanization, modernization and westernization. As a result there has been an increase in the GNP but at the same time the society is witnessing an increased materialistic concern and intense consumerism. The social and psychological changes in India since the achievement of political independence show economism as the overarching rule. Following utilitarian principles, developmental activities have been largely production oriented and aim at the greater availability of consumer goods. In this process, domination over nature and exploitation of natural resources have become prime concerns.

It may be noted that in the western thought, a liberalized individualist notion of human beings in line with the Judeo-Christian tradition is prominent. This mode of thinking assumes compatibility between individuality and communality, via the satisfaction of individual needs. This, however, has not been corroborated by our past experiences. The relationship between need satisfaction and social harmony depends on the availability and equitable use of resources, and the values and concerns of the people. The individuocentric bias recognizes society in its instrumental role, and moral considerations are frequently ignored. The resulting theories of persons are largely insensitive to the social world. The idea of a self-contained individual, however, cannot sustain life for a long time. It cannot respond to the demands of the post-industrial socioeconomic structures which require a globalized functional unit—a unit of constitutive persons who are not possessive, but consider themselves as guardians of social interests and assets.

Unfortunately the dominant form of psychological analysis is largely rooted in a mechanistic, deterministic and linear world view. It has a distinct focus on a self-contained and possessive individual at its core. Such an individual treats everything as commodity to be consumed for his/her prosperity and development. It reinforces destruction and exploitation of resources to satisfy the ever-growing personal needs. Within this framework, the individual's chief concern becomes seeking pleasure and avoiding pain. Things beyond the boundary of unique and separate self (e.g. environment, society and community) come into the picture only to the extent, that they inhibit or facilitate the pursuit of individual goals.

The practice of consumerism has continued to produce excess wastes which are beyond the earth's absorption capacity. The result is increasing illness, poverty, pollution, scarcity of drinking water and damage to the biosphere. The use of high technology and nuclear power have further multiplied these problems. All these happened within the modernist perspective which propounded human beings as the measure of everything. But this orientation is failing in its handling of the problems of the contemporary world. Interdependence, complementarity and sharing are increasingly becoming critical for meeting the challenges of survival.

Some Issues in Conceptualizing QL

The conceptualization of QL involves several issues, and a meaningful notion of QL can be obtained only by addressing these issues with a reasonable degree of success. Some of the pertinent issues are listed below.

Contextuality—The dominant orientation has been to adopt a common universal notion of QL. On the contrary, indicators of QL, by nature, have to be contextual and relative to a given socio-cultural setting. Thus what is considered as QL for one group or community may not hold true for the other. The eco-cultural setting of a community determines not only the resources and potentialities but also sets the limits on its accomplishments. The availability of resources is variable across groups and areas, and universal measures cannot do justice to what is needed across different groups and societies. A related question is about the interest of those who define and assess QL. They can be insiders (the people themselves) or outsiders (e.g. experts, powerful others). This will make a significant difference because the priority and salience of various factors are expected to vary across groups.

Dimensionality—The second issue is about identification of the dimensions of QL. There is increasing realization that QL is a concept which covers subjective as well as objective facets of experience. From a subjective perspective the experience of wellbeing is more important. This may include the feelings of optimism, hope, happiness, tranquility etc., which ensure satisfaction of needs. From this angle, greater the need-satisfaction, higher will be the QL. In contrast, the objective dimension is concerned with the investment of resources at local and global levels for achieving sustainable development. This dimension, therefore, is tied to more objective indicators such as the degree of security in economic and other areas such as, food, health, environment, community and politics. A comprehensive analysis of QL will have to address both the subjective as well as objective facets in these arenas.

Level of Analysis—Usually the analysis of QL is carried out with a focus on an individual's or group's conditions, in an isolated manner. This kind of micro level analysis remains incomplete as various systems and subsystems are interrelated. For instance, QL in the area of people's health depends on QL of the environment. A poor quality or unhygienic environment, becomes a threat to life. Since this universe consists of structures organized in the form of a nested hierarchy, taking things in isolation will not serve much purpose. A reasonably inclusive strategy is more desirable because coordination and support across different levels (e.g., individual, group, society) is a must. If something happening at one level has consequences that are negative for a higher or lower level, that cannot be considered as a contributor to QL. This has become a major challenge before many nations, as economic development through industrialization and urbanization is often accompanied by pollution, new diseases, alcoholism and the emergence of slums which become a threat to QL. Such experiences are quite common and indicate the need to reevaluate the exclusive strategy for improving QL.

Experiential Features—The psychological analysis of QL is largely focused on happiness, pleasure, contentment, joy and other positive emotional experiences. All these concepts are basically experiential in nature and represent fuzzy sets in terms of their coverage. They also involve value judgements. For example, if a person has achieved his goals according to his or her expectations and aspirations he may feel contented and report a high level of QL and *vice versa*. In fact, the experience of QL deals with a wide ranging phenomena having great depth. Contemporary research on subjective well-being does indicate empirical concerns, but has not been able to provide an unambiguous mapping of its territory. Also, its relationship with other variables is not very strong. Self-report measures of QL often depend on the mood of the person and various contemporaneous events.

Duration—The notion of QL in terms of subjective well-being may be looked at from an episodic perspective or viewed within a larger time frame. At an individual level, the moments of happiness during states of love, peak experiences, and *samadhi* usually do not continue for prolonged periods. However, the general life conditions at the individual, community or national level, also have enduring aspects, and therefore larger time frames, should also be taken into consideration for a comprehensive analysis.

Multifacetedness—The problem of QL is further complicated by the fact that life has multiple domains, and our experiences of satisfaction and happiness may take different forms. As a human being we have at least three major domains of functioning, i.e., biological, psychological and social. From the biological perspective the highest good is preservation of the species. The optimal functioning of this aspect of life ensures survival. At the social level, preservation of society becomes the prime concern. Thus, how others facilitate the functioning of social order is considered to be a contributor to the QL at the societal level. Within this conceptualization QL becomes synonymous with virtuous life. At the psychological level, the essential concern is self-enhancement. From this angle self-actualization is treated as a major motivational force for human beings, which may incorporate spiritual development.

The biological, social and psychological aspects of human life, however, may not necessarily be consonant. But QL in its totality is contingent on an optimal and harmonious functioning of all the three aspects. This entails adjustments, compromises and sacrifices at one or the other level. It may also be noted that greater emphasis on one part may result in conflict or lower level of QL in other areas. At an individual level, the magnitude of QL may be more or less dependent on one's capacities and the degree of support available to him or her.

Value—The notion of QL is intricately related to the question of values. Unfortunately in most of the analyses of QL, the question of values and morality is often reserved for the social aspect of functioning, and the biological and psychological aspects are treated separately. However, this scheme is inadequate, as societies precede individuals and constitute individuality. Also, the social order moderates and modifies the psychological and biological functioning of the people. So it is only through some broader value system that an adequate coordination among the three aspects of life can be achieved. Attainments in one area should not interfere with attainments in the other. At present, human pursuit in these three areas is often divorced from each other and excellence in the area of self-enhancement often goes against social welfare.

Conflict—The feeling of happiness, a major component of QL is often treated as tranquility—a state free from conflicts or struggles. This, however, does not seem to be the general case. The distribution of pleasures and pains in human life shows that for most of the people both are inevitable. The structure of our life is such that a worldly life totally free from pain, misery and wants is unrealistic and impractical. There is also the problem of defining pleasure and pain in absolute terms which perhaps is not possible. In fact, they can only be understood relative to each other. Therefore, happiness cannot be treated as free from conflict. This makes the nature of QL contingent on the appraisal of reality.

Towards a More Comprehensive View of QL

The Indian perspective on life is more comprehensive and articulates existential concerns in totality. In this tradition there is also a clear recognition of the fact that pleasure and absence of pain is one of the important goals of life. However, there is considerable debate over the question of whether pleasure or happiness is also good or worthy of being experienced. In this context, our attention is drawn to the distinction between what is desired (*Preyas*) and what is desirable (*Shreyas*). Thus, *Preyas* are considered as common pleasures and are treated inferior to the *Shreyas* which involve Supreme bliss and common good. The former is desired but the latter is desirable. This distinction places emphasis on the right or appropriate attitude. The experience of pleasure can also be looked at in terms of means and ends. If some source of pleasure is instrumental then it is considered to be of a lower degree than a pleasure which is valued for itself. Thus any pleasure is not inherently good and worthy of being sought. In fact, various ordinary pleasures are only apparently attractive. In reality they are illusory.

As noted earlier the social and psychological studies of QL usually take either the felt experience of happiness or satisfaction into account or look at the discrepancies between what a person has and what is his aspiration. At times comparisons are made in terms of a person's best past condition or with an ideal future condition. Both of these approaches focus on individual's motivational structure. Also individual-centeredness is dominant in the entire analysis. The motivational analysis of all types done in contemporary psychology aims at maximization of pleasures from the perspective of the individual. It is taken for granted that if an individual's material comfort increases, everything else will be all right.

It is obvious that this account of self-centered notion of QL fails to register the totality of our existential condition. It is quite possible that individual good may not be compatible with the social good and this has been the experience in many countries including India, where both poverty as well as affluence have accentuated simultaneously. The democratic-socialist political structure has in fact, facilitated the skewed distribution of wealth, favouring a few and marginalizing many.

The accumulation of material possessions and physical comforts is not linearly and positively related to happiness or life satisfaction. Perhaps, after a certain point they become negatively related or even unrelated. The self-contained individualism which is implicit in this view of existence, makes a clear division between self and other. It considers everything from the vantage point of an unconnected, bounded and separate self. Such a construal of self depicts an individual as transcending and dominating everything else and self distinguishes its unique identity from the rest of the world. The strategy implicit in this construction of self is to conquer the world by controlling it through various means.

Many non-western cultures conceptualize a relational self which does not draw a sharp line between self and other. Self is embedded in the group and is constituted by others. This kind of extended or ensembled self-construal is quite broad and to focuses our attention on the totality. Self and others share a common life space. This kind of reflection allows for a decentered view of self, wherein there is shared responsibility for self as well as other. Perhaps, the question of QL should be dealt with at the level of this shared reality of life in which life in itself becomes a value and quality. Within this perspective QL will become a broader concept that views life in totality and not just in terms of the experience of one single person. Such a conceptualization will serve to eliminate the conflict which often emerges due to partitioning of self and other and giving priority to self at the cost of others.

Perhaps, the causes of poor quality of life or lack of happiness and increased stresses are located in individual's needs and external circumstances. There is a strong self-serving bias in attribution of success outcomes. Therefore, failure is often externalized and attributed to factors such as luck, fate, task difficulty etc. One does not like to accept the responsibilities for one's failures because this involves sharing the burden of pain or suffering. In the Indian tradition on the other hand, suffering or *Dukkh* is an inevitable part of life and is always present. Life is structured in such a way, that from birth to death, pain, misery and suffering are continuously experienced. The causes of suffering are considered to be rooted in desires and various other afflictions such as egoism, ignorance, attraction, repulsion etc. The solution is found in management through control and reduction of desires, their modification, and inculcating a tolerant and balanced attitude. The Indian view emphasizes self-management rather than control of external environment. Unfortunately in contemporary discourse this perspective has been totally ignored and is regarded as an ineffective strategy of self-management. It is conceptualized as secondary control or defense. However, if we appreciate the viability and holism of a transpersonal perspective and do not subscribe to strict determinism, this may facilitate a shared and relational view of reality.

Conclusion

In conclusion it may be stated that the individual-centered perspective on QL is insufficient as it ignores the larger context in which the individual is located and by which he/she is constituted. The contemporary challenge before humanity in terms of the problem of environmental degradation, inequitable distribution of wealth, unemployment, diseases and poverty, pose a serious threat to the peace of mind, contentment, happiness and satisfaction. The idea of modernization and development has failed to realize the goals of human welfare and in sustaining the quality of our habitat. The threat to QL is now perceived at a global level and unfortunately the poor countries are suffering more and have to bear the burden of the rich and powerful. Conflict and chaos have become inevitable parts of our experience and much of it owes to our own design. For a better future and to sustain life processes, quality needs to be added to human goals and concerns. Instead of looking at the reality from a narrow personal perspective, there is a need to expand our consciousness. A dispassionate analysis indicates the limits of the consumeristic notion of QL. The dream of a materially comfortable life has not been realized. This style of life rooted in the *having mode of existence* indicates a crisis of values. We, the occupants of this planet earth, are not only consumers but it's trustees as well. We have to contribute to the solution of problems which are largely created by none other than ourselves. It is high time that we realize that there is a limit to growth and it is not the quantity of material possessions, but the quality of existence which makes life worthwhile. Quality of life is rooted in recognizing interdependence, complementarity and sharing of existence. This requires critical self-enquiry and reflection on our part. Living in accordance with an integrated and holistic view of the universe is the only way out from the present crisis. The choice is ours and clearly it should be in favour of respecting life—the highest quality one can have.

8

Contemporary Human Condition
Challenges and Possibilities

Human life in the contemporary world is full of conflicts and contradictions, at different levels of existence. Increasingly, the tendency to intervene and destroy the natural processes is assuming alarming proportions, and societies are rapidly moving towards a kind of irreversible artificial world. The mecha-tronic technology of this age is leading towards structuring a new form of reality and discontinuities in patterns of lifestyle appear to be remarkable. In the developed countries, the relationship between knowledge, technology and environment characterised by information economy is undergoing large-scale transformation. These challenges are being addressed by two visions for future. One considers a new ecological paradigm which intends to impose potent physical and biological restraints on human affairs. The other alternative largely shared by the optimists suggests that human beings must assume responsibility for governing evolution and, therefore, plan things accordingly by developing higher order technologies.

In the context of sociopolitical systems, the 20th century witnessed failures in different degrees of two great experiments, capitalism and communism, and their derivatives (Steenbergen, 1990). All of them have been trying to achieve the goals of growth and equality through various means (e.g., market, plan tax, zerosome game). They have achieved different degrees of freedom, security, welfare, identity and well-being for the people. At present, they are demonstrating crises and dangers of inequality, and social unrest. It is deeply felt that existing paradigms are unable to address the issues of economic dynamism, social equity and ecological stability. The idea of progress and the dream of unlimited growth and development have proved untenable.

Developmental Experience

Life in the so-called developed part of the world apparently presents a picture of greater material wealth, high literacy rate, high rate of consumption and reduction in the duration of working hours. However, it also presents a scenario of suicide, alcoholism and other symptoms of social pathology and mental disturbance. The dream of a materially comfortable life has not been realised if one considers the signs of mental imbalance. Taken together, these features characterise a 'having' mode of existence. A critical appreciation of the prevailing situation implicates that there is something wrong with the basics of this style or mode of life. In fact, western civilisation is documenting a crisis of values which define human existence. The developing and underdeveloped countries of the world are also undergoing crises, but in a different way. Most of the nations, earlier grouped as 'Third World' countries, are poor, and are in a process of decolonisation. The experience of colonialism has resulted in a culture which is predominantly derivative and imitative of the West in its spirit. The processes of globalization, privatization and liberalization are shaping the world. These forces have contributed in GNP, but at the same time have made mean materialism, valuelessness and mindless consumerism central to the contemporary social milieu.

The psychological change that has occurred since the achievement of political independence in India has resulted in a high degree of insecurity, lack of faith, cultural illiteracy, and adoption of values and behaviours which are culturally incongruent and constitute a worldview which is dysfunctional in a longterm perspective. Our identity is seriously and persistently threatended to be diminished and lost. Our developmental plans and efforts were largely organised within the framework of economies. Following utilitarian principles, the same have been production oriented and aimed at greater availability of consumption goods. Domination over nature and discovering new ways of exploitation of natural resources became priorities. Under the impact of the western model of modernisation, cultural strengths and indigenous knowledge and knowhow were viewed negatively and declared as impediments in the path of development.

Change in Scientific Paradigm

Newtonian physics in which every event was determined and the 'world' was atemporal, had no scope for chance and probability. This kind of mechanistic model provided a limited and distorted view of reality. In the early 20th century, Einstein put the observer back into the system and subsequently, the quantum people recognised the significance of indeterminancy. Attempts at understanding the biological and social systems within a mechanistic framework could not grasp the complexity of self-organising systems which often exhibit 'order out of chaos' by the processes of bifurcation and rearrangement (Pregogine & Stengers, 1984). In the context of these developments, the transparent view of a physical universe constructed within classical science became insufficient. It has been realized that instability and fluctuation are significant features of complex systems. These features demand change in the construal of the relationship between human beings and nature.

A new perspective on man, society and the world is gradually emerging. In social sciences, too, we note a new swing towards 'holism' which rejects the prevailing mode of activity characterised by specialisation, fragmentation, and atomistic analysis. For a long time, Descartes and Newton served as the models, and psychologists adopted this physical science paradigm and continued with the empirical analytical approach. They remained concerned with prediction of future behaviour (and at times reproduction of past behaviours) (see Harre, Clark & Decaro, 1985). The new holistic view proposes that : (a) the parts can only be understood through the dynamics of the whole; (b) processes are more crucial than structures, and (c) Cartesian dualities are untenable. The holistic perspective is not in line with habits of psychologists formed during disciplinary socialisation under the forces of modernism. It also rejects the anthropocentric view. Instead of emphasis on subordination, hierarchy, struggle for life and survival of the fittest, the holistic view recognises interdependence and a dynamic form of cooperation. This is clearly evident in the case of self-organising systems which involve autonomy, self reference, spontaneous development and sharing of learning capacity.

Limited Psychological Perspective

In the Western world, psychology emerged with a focus on liberalised individualist assumption in the Judaeo-Christian tradition. In doing so, a compatability between individuality and sociality *via* the satisfaction of needs was postulated. This, however, has not been substantiated through our experience. The linkage between need satisfaction and social harmony has proved to be a myth. The anthropocentric and individualistic bias had recognised society's instrumental role, and moral considerations were largely disregarded. The resulting theories of persons were insensitive to the social world. They however do not go well with the demands of the post-industrial social reality. This demands a globalised functional unit of constitutive individuals who are not possessive, but treat themselves as guardians of social interests and properties (Sampson, 1989).

In psychological science, human behaviour is often explained as a variable to be explained in terms of a set of antecedent and mediating factors, including individual dispositions and other psychological characteristics, on the one hand, and environmental features, objective as well as perceived, on the other. This partitive analysis is rooted in a mechanistic, deterministic and linear world view with a distinct focus on self-contained and possessive individual as its core. It transforms ecology and environment into a commodity to be consumed by human beings to achieve prosperity and development. It denies the possibility of perceiving human beings as part of the total existence or *sristi,* and reinforces destruction and exploitation to satisfy the growing needs of the individuals. The situation is usually depicted as one of competition between human beings and nature in which human beings have to assert for their supremacy.

The individual-centred vision of mainstream psychology inevitably situates all psychological processes in the individual, and the person is depicted as a hero whose single concern is seeking pleasure and avoiding pain. Thus, a person has obligations to himself or herself alone. Environment, society or community come into the picture only to the extent they inhibit or facilitate individual functioning. This perspective cannot go too far in handling the problems of the contemporary world. Complementarity and sharing are increasingly becoming critical to survival across the globe. Taking this aspect of existential reality, demands decentring of thinking, and reformulating the mode of conducting psychology and reordering its priorities. Until recently, the dominant paradigm in psychology was characterised by a microscopic, reductionistic, and physicalistic view. Now there is a need to evolve a holistic ecological paradigm for psychology which may address the complexities of human behaviour. This can be achieved by focusing attention on the relatedness and constitutive nature of human existence. The organic world-view of indigenous Indian thought which assumes correspondence between macrocosm and microcosm and recognises a basic sharing of all life forms, has immense potential to contribute to this end.

Environmental Challenges and their Management

The contemporary strategies of managing environment are largely guided by the underlying construals of human environment interface. The scientific technological world-view maintained a mechanistic stance and considered environment as a resource to be exploited and consumed by human beings. The ideology of modernisation and development had no consideration for the carrying capacity of the planet earth. This has resulted in an imbalance of the ecosystem. The practice of consumerist lifestyle continued to produce excess wastes that go beyond the earth's absorption capacity. The developing countries are being threatened by illness, poverty, pollution, scarcity of drinking water and damage to the biosphere. The use of high technology and nuclear power have made many problems global. All this happened within the modernist perspective which propounded that humans were not only the central players but masters of this universe.

An alternative world-view which is rooted in the Indian and many other cultures emphasises the relatedness of human beings with other forms of life and things constituting the universe. All life forms are constituted by the same set of elements. This view recognises interdependence and connectedness, and considers existence of everything in a relational framework. It is proposed that nothing exists in isolation and, therefore, behaviours of all the entities inhabiting the universe are interlinked. This kind of cosmocentric vision emphasises dialogue and co-existence with the surroundings. Human life is viewed as enveloped by a dynamic cosmic order. Projecting an integral view of man, nature and universe, it stresses on the mutual reinforcement of human and nonhuman beings. Recent thinking is recognising the intergrated and holistic view of the universe. The ecological wisdom, which propounded that humans and the rest of creation are partners, is being rediscovered. For instance, the GAIA hypothesis views earth as a single living organism.

The emergency currently faced by humankind demands urgent attention and intervention at local, national and international levels for conservation and maintenance of the entire ecosystem. To this end, multidisciplinary or crossdisciplinary approaches are warranted. Behavioural and social scientists can contribute to the problems of environmental management by (a) analysing the impact of environmental changes on human beings; (b) developing appropriate strategies to deal with environmental stresses, and (c) by developing awareness and changing attitudes and behaviours of people in favour of a viable and balanced ecosystem. Success in this endeavour will be attained only if we are able to promote an attitude of reverence for life, encourage respect for sustainability and plan within the carrying capacity.

It seems that a major error has been that the nature of human existence was misconstrued. Transcending nature through self-awareness, reason and imagination have often been treated as factors which define human existence and distinguish it from animal life. The human needs that emerge from the objective condition include relatedness, transcendence, rottenness, sense of identity and a frame of orientation and devotion (Fromm, 1976). Even this view has a number of limitations as it gives greater emphasis on human as the center of universe. The springs of action are in the person, and other forms of life assume a secondary place.

Towards a Better Future

At the beginning of the twenty-first century, the human condition does not show a distinct improvement over the past, concentration of resources and power in the developed countries and abject poverty in the masses of the undeveloped countries, present a threat to global peace and societal development. Gradually, all problems are assuming a global nature (*e.g.*, pollution, atmospheric changes, demographic patterns), and their impact is not localized (Dube, 1988). In order to save the common future, changes in the world order are necessary. To this end, interdependence, mutual sharing, and a constitutive framework of sociopolitical functioning is warranted which may provide a guarantee for maintaining a harmonious relationship between cultural pecularities and universals.

Achieving a 'being mode of existence' demands going beyond the anthropocentric view. In this context, traditional Indian values offer a perspective which has relevance to the current crises. The view of 'man in relation with rest of the world (or universe)' and not as a discrete organism, self-control (swadhinata), continuity across life forms, interdependence, creativity, transcendence of time and space, togetherness, and emphasis on totality of life signify the processes that are not only of historical importance or mere hypothetical notions, but are pregnant with meanings which point towards a model of 'man as part and parcel of the whole of Nature'. Man is certainly not the measure of all things. In this view, 'nothing stands isolated' and 'everything has its repercussions in a wider sphere of time and space'. These ideas imply a more relevant and viable perspective on human functioning. Interestingly, these constituents of the model of man are in line with realisations of modern science. For instance, exchange of matter and energy, indeterminancy and complementarity are now considered as important principles in physics. Realising relevant social change, necessitates a creative response on the part of psychologists as well as other social scientists with an open mind and from an indigenous perspective.

9

Deprivation Research and its Policy Implications

Indian society evinces wide gaps and disparities along socio-cultural experiences, economic sufficiency, residential conditions, and educational experiences. This situation provides a psychedalic array of hues with vast differences in experiential background which makes deprivation an important problem. The interest of Indian researchers in 'deprivation' however owes largely to the influences of sensory deprivation research and studies of cultural deprivation in the United States. The increased concern with the weaker sections of society such as scheduled caste (SC), scheduled tribe (ST), minority groups and the people below poverty line has been another impetus to this end. The focus of psychological research has been on mapping the differentials in cognitive, affective and personality domains of psychological functioning of the deprived people.

Literally the term 'deprivation' means the process of dispossession. In experimental studies the operation of deprivation assumes reduction of sensory inputs available to an individual which can take different forms depending upon the number of sense modalities involved. In current usage it is used to refer to the lack of (necessary) experiental base considered essential for healthy growth and development.

However, despite considerable effort it is yet not precisely known as to what are the necessary and sufficient conditions, facilitators, and inhibitors of human development. Also, the way cultures shape individuals is a long term, complex and intriguing process. While these gaps in our knowledge have to be taken as stimuli for future research rather than constraints, a meaningful analysis of deprivation cannot ignore the issue of development and its context. So far, this issue has received only limited attention. Human development is shaped in the context of intimately interacting different subsystems of environment arranged in a nested heirarchy. Recent studies document that very young children are also equipped with sharp and effective cognitive apparatus. However, owing to perference for individualistic and micro approach, the researchers have largely taken the position that deprivation is an individual level phenomenon or is the property of an individual. The tragedy is that those who have viewed deprivation as an environmental phenomenon too have reduced it to the level of the individual. This reduction has shifted the focus of intervention from environment to the individual. As a result instead of enriching the deprived environment and addressing the structure that maintains it, efforts are made to change the deprived ones through training, motivating, and changing attitudes which are bound to create greater frustration as the raised expectations are not congruent with the existing environment. The lack of support from the environment further accentuates the problems. The environment is maintained and nurtured by some features of the traditional social structure and ineffective implementation and monitoring of intervention programmes.

Disparities and inequalities in opportunity are present in varying degrees in different societies resulting in the formation of advantaged/privileged and disadvantaged/underprivileged sections of society. These extremes, however, are not always well articulated and to a large extent remain undifferentiated. The specific composition is usually linked to the particular socio-cultural ethos and historical political context in which they emerged and the contemporary social reality, in which they exist. Thus the nature, magnitude, and extent of deprivation which distinguishes various group is inherently relative in nature. As such it can be asserted that the same can reasonably be assessed with the help of objective indices of the prevailing life conditions. Of course, that again would be relative since the standards of life and distribution of diverse resources widely differ not only from country to country but also from one region to the other in the same country. That is why the Black-white dichotomy cannot be imported and extended meaningfully to depict the contrasts between Harijans vs. Brahmins, urban vs. rural, tribal vs. non-tribal, and so on and so forth. In this context it may be noted that the distribution of income in society as indicated by the per cent of population below poverty line, features of vertically organized social structure, ecological variations and availability of physical resources in India present a contrast to the western world. The western categories of deprived-nondeprived are not congruent with the Indian social reality.

The relative nature of deprivation is also expressed in the feeling of deprivation for which some researchers have coined the term *relative deprivation.* This refers to the experience of being deprived rather than the objective state of deprivation. For instance the lack of necessary physical support yields varying degrees of discomfort in different individuals. Conceptually it should be above the threshold (adaptation) level if it is to be considered deprivational. Such a threshold would, of course, depend upon the comparison level opted or used by the individual/group in question. Thus one feels deprived as compared with other individual/group which may or may not be related to what he or the group possesses. This implies that deprivation is a subjective construction rather than a given property of the external, physical and social environment.

The researchers have coined several terms that are used independently as well as interchangeably to refer to the different facets of deprivation. Some of these terms are value loaded. In the contemporary period when search for alternatives and pluralism in thought are valued it is not adequate to label a given culture/group/individual as deprived simply because it does not conform to the chosen set of exogenous standards. The indiscrete use of phrases like 'culturally deprived' or 'disadvantaged' show biases and preferences of values rather than the experience of deprivation.

To this end three major strategies of operationalizing and measuring deprivation have been adopted. They distinguish between different degrees of deprivation in terms of *environmental categories* (e.g., residential background, institutionalization), *social categories* (*e.g.* membership of special cultural, economic, ethnic or caste groups) and *experiential units* (*e.g.* indices of deprivation in terms of environmental inputs and individual's experiences). While all the three approaches consider deprivation as the deficient environmental condition, the first two categories involve only limited characterization of these deficiencies. The third one makes it explicit and takes into consideration, the extent of within category differences in the degree of deprivation.

A close perusal of the uses of the term deprivation suggests that it consists of deficient, substandard, and inadequate environmental conditions which interfere with the development and growth of individuals. An environmental condition is deprived to the extent it adversely influences individual's attainment across various domains. Unfortunately, the focus of attention has shifted *from environmental deprivation to deprived persons*. The property of environment was attributed to the individual. This point needs elaboration since it has policy implications. Once we consider that the deprived persons are behaving in a given manner then the strategy of intervention will be directed towards the individual rather than the environment. Perhaps we need a two-pronged strategy aimed at changing the individual and bringing changes in the environment.

In the last two decades considerable data have accumulated pertaining to the relationship between deprivation and psychological functions in Indian context. It is, however, difficult to draw unequivocal conclusions as there are variations in the criteria of deprivation, type and quality of measures used and regional variations. The efforts to understand the consequences of deprivation have so far largely been confined to the demonstration of differences across groups formed along variables such as caste, region, SES, and sub culture designated as deprived and privileged. The differences that have been uncovered in the areas of cognitive functions, intelligence, affective and personality functions, mental health, attitudes and academic attainment are not unequivocal and are also fraught with several methodological limitations. Their comparability in terms of sampling and appropriateness of dependent measures often complicate our understanding of deprivational effects. Notwithstanding these limitations some trends can be deciphered (See for details Misra, 1983; 1990; Misra & Tripathi, 2004; Mohanty & Misra, 2002; Rath, Dash & Dash, 1979; Sinha, Tripathi & Misra, 1982; Singh, 1978; Zafar & Alam, 1985) These trends may be briefly summarized as follows:

The studies have generally shown a negative effect of deprivation on perceptual, conceptual, Piagetian, problem solving and representational tasks. Long term memory, recognition and learning also evince similar trends. However, the measures of short term and incidental memory do not show such effects. Linguistic achievement and syntactic ability show significant adverse effect of deprivation. Cognitive styles also differ and so does the performance on scholastic and academic achievement measures. Also, some positive characteristics like boredom tolerance, vigilance, and drawing ability are found in the disadvantaged children.

The motivational characteristics of deprived children include low achievement motivation, low aspiration, low expectation, neuroticism, insecurity, rigidity, conformity, anxiety, extra version, alienation, low self image, past orientation, lack of future orientation, less positive self image, external attribution for success and internal attributions for failure, and use of dysfunctional coping strategies like escape, helplessness, hopelessness and denial. The effects have largely been found to show cumulative deficit though it cannot be considered as a general rule.

In general, the deprived people perform on majority of the cognitive and intellective tasks at a lower level than their priviledged counterparts. However, on the measures of competence and ability to learn the discrepancy between the two groups is found to be low. The deprived and non-deprived groups do not evince substantial difference on cognitive and behavioural measures during early years. However, the gap between the deprived and non-deprived groups increases with advancing age. This indicates the role of non-intellective factors in accentuating the differentials between the two groups (*i.e.* school climate, teacher expectations, parental support etc.) which often lead to cumulative deficit.

Viewed against the mainstream the deprived groups lack

those qualities, dispositions, skills, and values which are considered necessary to bring success and upward mobility in the contemporary competitive world. They show low persistence and tend to leave the field. They are not equipped with the necessary skills nor prepared to cope with the demands of society. The goals aspired by them are often beyond their reach.

The picture which emerges from these differentials suggests that the deprived or disadvantaged children and young adults display a less efficient functioning than their advantaged counterparts. Such differences increase with the complexity of task demands and advancing age. The explanations of deprivational effect put forward, favour a *difference* or *deficit* explanation or a combination of the two. These explanations have different degrees of provisions for change in behaviour. The difference hypothesis attributes performance differences to inadequate measures and procedures, and considers that the differences are artificial and none of the groups is more or less efficient. They do not differ in basic processes. The deficit hypothesis assumes that the deprived are deficient in their cognitive, motivational and personality characteristics. In its extreme form it considers that the differences reflects genetic differences. In its mild form it assumes that the differences reflect lack of cognitive structures or skills owing to lack of necessary environmental support. The third possibility subscribes to the view that none of the two explains the situation adequately. Misra and Tripathi (1980) have contended that the performance differences involve deficit as well as difference depending upon the nature of skill and domain assessed.

Deprivation is primarily an environmental phenomenon. It stands for a situation in which alternatives for human growth and development are reduced and people are subjected to inhuman conditions, which generate, sustain and promote a noncontingent relationship between action and outcome. As Misra (1988) has argued, they gradually internalize the view that their behaviour, and therefore, their existence, has no relevance to the broader society. The resultant feelings of helplessness and apathy restrict them from undertaking significant steps in the direction of increasing self efficacy.

Evidence in support of this contention comes from studies on achievement attributions (Jain & Mal, 1984; Mal, Jain & Yadav, 1990; Misra, 1985; 1988; Misra & Misra, 1986; Misra & Jain, 1988a; Singh & Nathawat, 1989; Singh & Misra, 1985, 1988). The experience of helplessness contains the belief that one's actions are dysfunctional and environment is non-supportive. These beliefs ultimately lead to an expectancy frame which is substantiated by their experiences in various spheres of life. For instance Sharma and Tripathi (1988) have reported that "teachers credited high caste students for their success but did not blame them for failure. While scheduled caste students were held responsible for their failure but were not given credit for their success. Attributions which teachers made became instrumental in sustaining their expectations and possibly, were communicated to the students. Since there is evidence to suggest that high caste students perform better in schools, whereas scheduled caste students do not do well, it seems teachers' expectations and attributions could be responsible for their performance" (p. 55). The outcome may result in diverse reactions such as withdrawal, resignation, confrontation, hostility, and alienation. Which one will occur is determined by the situational factors as well as an individual's skills and dispositions. In this context, the state provisions of loan, subsidy, and other facilities are expected to play an important role. It is generally held that these programmes were not effective in changing the prevailing state of affairs. Rather, they have made the feelings of dependence and inadequacy more strong. The chain of reactions continues to maintain the prevailing situations.

The only way out is to involve the deprived people in such a manner that they should feel empowered and experience that their behaviour is functional, productive and they can influence the environment. There are some examples in our social life which illustrate that this strategy is successful. We find that the provision of resources alone does not guarantee social uplift, unless it is accepted and properly utilized. Also, it is not necessary to provide a highly enriched package. Deprivation is curvilinearly related to performance and motivation and, therefore, certain degree of deprivation may yield facilitating effects. Finally, it may be pointed out that the problem of deprivation should also be examined at the societal level. Instead of blaming the victims we need concerted efforts towards identifying the structure and components of environment at individual and group levels which make the people deprived, and evolving suitable means to make the deprived people feel competent and self-reliant.

It can easily be discerned that different policy implications involving change in the individual or change in the environment or change in both can be initiated. However, a moments' reflection would make it clear that change in the person or in the environment alone would not be sufficient. Since both constitute a system, change in one has to be supported by change in the other. Failure in co-ordinating the two will add to the problem. The coordination of the two components of the system is needed. Enrichment programme, attitude change and motivational training with a focus on individual in the absence of required environmental support, would not succeed.

Assuming that the preceding generalizations are true and we are sincerely concerned with the betterment of deprived people, a two-pronged strategy is warranted. First, we have to take a clear stand in regard to the goals to be achieved. Second, we have to plan for necessary steps for achieving those goals. Unfortunately, psychology has largely been focusing on contextualized individual and proposing changes mainly in the individual and not in the socio-economic situation. The unfavourable experiences in one's life are attributed to faulty mechanisms within the person. As Prilleltensky (1989) has argued psychology has a dual responsibility in this context. On the one hand, it has to develop awareness of the prevailing circumstances and ideology, and on the other hand, it has to provide a version of possible framework for human welfare. These processes are known as conscientization and annunciation. Conscientization refers to the process whereby people attain an illuminating awareness of not only the critical environmental conditions that are shaping life but also of their capacity to transform that reality. Annunciation is viewed as the act of convincing a just social arrangement in which the well-being of the population is fostered. These challenges need to be responded to in a creative fashion.

10

Continuity and Change in the Indian Family System

The study of family in the Indian context is a complex enterprise and making generalisations about the Indian family is hazardous. The reasons for this state of affairs is twofold. *First*, this institution is situated in diverse cultural traditions which define the structure and functioning of family in different ways. *Second*, contemporary influences from economic, educational, political and cultural forces are directing the patterning of families in several ways. As a result we notice considerable variation in the configuration and functioning of Indian families which evince the elements of continuity as well as change. What follows is a somewhat informal analysis of this situation based on personal experiences and reflections.

The Institution

The institution of family in India has traditionally served as the basic unit of human social life. The socially shared notion of a person, as opposed to the notion of a person as a separate entity having clearly delineated boundaries, has been the Indian ideal, and various social institutions and practices have grown to nourish and maintain harmony between the twin goals of individuality and relatedness. Family has proved to be a strong and powerful instrument to this end. In addition to the provision for necessary support for survival, family also worked as a buffer and helped in maintaining an optimal level of stress and coping. By providing mutual care and regard to its members, family also extended the scope for stability and development and allowed realisation of various personal and social goals.

Families however, are dynamic organisations. Like open systems they are reciprocally related to the social environment. Also, like living organisms they grow and change. The nature and structure of family varies from the perspectives of it's members too. Thus while it is located in the same space for men during their entire life span it changes for women after marriage except in matrilineal tradition. There is considerable variety in the kinship relationship. In fact the kinship universe in the Indian society is very large. All families follow a cyclical pattern in their life span. In the beginning, families are joint, then they split and form nuclear families which grow and follow the same cycle. The structure, composition and roles in families often undergo important changes. The broad-spectrum social changes also influence families. For instance, the age of marriage of girls is increasing. Women are asserting their rights. The generation gap is widening. Competition is increasing and conformity is decreasing. The mass media is playing a crucial role in shaping the desires and ideals of the people. All these factors are introducing new elements in the family dynamics.

Family as a Value: The Indian Tradition

Indian family system, however, still maintains certain distinctive characteristics. Instead of being a deliberate arrangement on the part of a single individual or a person, it operates on a continuous basis to nurture the relationships, interdependencies and shared goals. As a result, people's identities, self-esteem and ideas about worthiness are often conceived and organised around the notion of family. In this way family constitutes a value. This is reflected in the development of what Rolland (1988) has described "we self regard" and "familial self". This kind of expanded self is not only open to others but is constituted by the significant others. Such a person endeavours to act as a part or element of the whole. As such, his major task happens to fit into the social network of family. He or she is embedded into it and owes existence to the shared reality. This view is in sharp contrast to the view of separate, independent and autonomous view of liberalised self popular in most of the industrially well-developed western cultures. In fact these two traditions have articulated the self-society relationship in two different ways.

The Indian family-orientation maintains the social bond, facilitates group goals, enhances cohesiveness among the group members by promoting sharing, cooperative spirit and interdependence. This is achieved through a hierarchically organised authority structure in which individual differences in motivation, aspiration and goals are ignored to a considerable extent particularly when they appear as threats to the structure of group goals. This kind of scenario often prevails in the middle class families in which decision making is guided by a dominant concern for the overall care of the family as a single unit. This necessarily requires appropriation of individual's wishes and needs. The extent of appropriation varies across different domains. Thus marriage and job are treated as family matters whereas pursuit of knowledge and religious activities form a domain with greater scope for individual freedom. The Indian village clearly projects such a situation.

Since long, the functioning of family as an organisation has been situated in a nested hierarchy in which family, caste group or community and village were intimately related to each other. The families were called *kutumb* and *parivar* and often consisted of individuals from several generations. Known as joint families this kind of familial configuration is taken as the natural form of family. In such a setting, growth, development and various activities are organised with peculiarly societal or communitarian focus. Individuals are shaped within the framework defined by this focus and the calender of life events is designed and programmed to cater to the needs of family. Apparently this system did operate reasonably well for quite a long time as far as social harmony and cohesiveness was concerned. However, with increase in social mobility, western education, migration to cities, urbanisation and industrialisation, this system is suffering a major setback.

The Changing Scenario

Economic pressure, growing needs of individual family members, consumerism, destruction of rural-agricultural base and growth in size is increasingly making the sustainability of large size families more and more difficult. The social, economic and emotional costs of such pattern of family life is making them unviable. As a result, the members of (joint) families are distantly located at different places. The structure of these family units is becoming *nuclear* which consists of wife, husband and children. These units, however, frequently maintain relationships with members of horizontally and vertically extended families on maternal as well as paternal sides. The strength of these relationships depends upon physical proximity, social status and degree of contact with villages. For instance, families having agricultural land and other properties, which are sources of incôme, have stronger contacts with their relations and village community than those who have lost their roots in villages (or small towns). It may be noted that the links with the extended families are occasionally expressed in various rituals and ceremonies (*e.g. Upnayan*). Arranged marriage, which is still the most frequent form of marriage, happens to be a matter decided at the level of family. Of course, increasingly the consent of brides and bridegrooms is also being sought.

Thus, what we notice is that for a vast majority of Indians, family continues to be the greatest concern. However, individual desires and wishes are emerging very strongly and a tension is being experienced between these two orientations. These tensions are closely related to the loss of power of the institutions of religion, caste and kinship. New individual-centred and somewhat secular institutions are emerging. This situation is leading to discontinuities of various types. The attitudes, values and beliefs, once held in high esteem, are being questioned and reprioritised. Concomitant to this, changes are visible in commitments, identities and relationships of the people. For want of space a detailed analysis of these issues is beyond the scope of this essay. However, it may be pointed out that freedom from caste and family based occupations, women's education and urge for personal accomplishment constitute as most crucial factors determining the family dynamics. With increasing emphasis on individuation and autonomy, modern parents are deliberately trying to shape their children's lives in a different way. Stricter child rearing coupled with greater expectations on the part of parents is creating an unprecedented demand on children. Also, with increase in the number of employed mothers the ecology of family is undergoing significant change. The interaction pattern, its temporal organisation and quality among the members of dual career families evince considerable shift from the earlier pattern. Its effect partly depends on the nature of supplementary and/or substitute child care and nature of mother's job.

The Challenge Ahead

The changes in family scenario in Indian context are pointing towards increasing instrumentality of the family system. This is an inevitable outcome of heavy competition for procuring the resources, high degree of individualism and growing concern with material gains. One's identity with family is becoming utilitarian. In other words people are engaging themselves in social relations in an exchange framework in which perceived rewards and costs are decisive. The breakdown of joint family is also partly related to this kind of change. Increasing emphasis on rights over duties, expectations over obligations and personal concern over the concern for family are contributing towards high degree of intra and interpersonal conflicts in urban middle class families. Such conflicts take different forms in the life span of a family. Being developmental in its constitution the composition of family varies across time. It cannot be considered invariant or static. Thus, daughters or sons remain as such in relation to their parents throughout their lives but they also grow and through various stages develop their own families. In this way he/she has to (a) maintain relationships with parents and (b) perform the functions of an independent person. This reflects the tension between *being* and *becoming*. Of course it is experienced differently by sons and daughters during their life courses. The fact, however, remains that such conflicts are major threats that demand innovative solutions on the part of family members. This is particularly important in the context of destabilising influences of contemporary culture which stresses on immediacy, functional (or utilitarian) values, practical considerations, comfort and pleasure. Having market as its basic model or root metaphor everything is conceptualised in terms of (monetary) exchange organised on short-term basis and people are viewed as commodities or (human) resources. All this is dehumanising and disturbing not only because they are not in conformity with the Indian tradition but also due to the fact that they have not proved to be effective solutions in Western societies. It is, therefore, necessary to strengthen the complementarily and interdependence of individual and family or community. Indigenous constructive effort to this end is warranted for enhancing the quality of human social life.

It may however, be observed that family system in India, particularly, in the urban educated class, is undergoing transformation. The economic, social and cultural forces are reinforcing individualism and changes intrinsic to the traditional family system are leading to structural changes. They are questioning the roles, obligations and relationships which have been nourishing the inherently social texture of the Indian family. The tensions experienced in the course of pursuing personal (individual) goals and familial goals are becoming an important source of stress. This situation calls for a perspective which views individuality and relatedness as mutually constitutive and complementary. Such a solution shall constructively utilise indigenous values in which family happens to be a value rather than a mere instrument for personal growth.

11

Education and the Challenge of Cohesion

One of the major goals of modern education is to contribute to the development of responsible citizens of a democratic nation. This necessitates sharing, trust and belonging as desirable dispositions to be inculcated among the students. Each student, however, possesses aspirations and natural tendencies unique to itself. If education aims only at bringing out the best in him or her as a student, then it will fail to achieve the goal of socializing. Education must attend to the conflict between individual and society. It should also see that not only the personality of persons is nurtured but training is imparted in resolving such conflicts so that a constructive, peaceful and productive national/social life becomes possible. Education is a must for human dignity and respect.

From a societal perspective education is a social instrument or design to fulfil social purposes. In this context, a society ensures its existence through the processes and institutions of education. A teacher, represents society and works as a part in the chain of cultural communication. It is his or her responsibility to make a social/moral individual. It is through him or her that a society constructs people of its imagination. This implies a close relationship between education and social development. According to traditional theory a large segment of population must become literate in order to become a participant in the process of development. Education is now treated as an investment in human resources so that new ideas and a better preparation for development can take place and ability and genius can be utilized for developing the society.

The system of education, however, is a part of the broader social system. Therefore it should not be seen separately or in isolation. Indian higher education started under the British colonial rule. It had to represent and preserve the British interest. It did not help the common man to be educated. Nehru, the first Prime Minister of free India did invite for revolutionary changes in the basis of education. However, economic development, modernization, social reform, social reconstruction and economic progress occupied the main place on the agenda of planned social change. In this process, expansion of educational facilities, literacy and free primary education were emphasized.

The Indian Constitution is committed to the right for education for all. There has been development in enrolment at all levels. However, illiteracy is still very much there and educational standards are deteriorating. Regional disparity in educational opportunity is prevalent. The Indian education system has remained traditional and socially less sensitive. It is not taking the shape of a force capable of evolving meaningful social activity necessary for the development of a healthy and a just society. The same, however, has not received significant attention in the process of educational planning in India.

From a historical perspective, India has been a country where á confluence of a variety of cultural and social groups has been taking place. People of different regions were allowed to pursue their own distinct styles of living. They were not converted to one style. They lived side by side. The notion of '*Vasudhaiva Kutumbakam*'—was operating. The different groups were allowed to practice their individuality in a shared social space without much conflict.

At present the Indian society is marked by its wide range of diversity in terms of race, religion, language, etc. The co-existence of groups could maintain their identity in such a way, that a few decades ago a Hindu orthodox Brahmin family—head could trust his Muslim neighbour to the extent of keeping a box full of important treasured objects at the neighbour's house during his own village visits with the family. Such a feeling of trust which existed even between diverse groups seems to be missing today. An endeavour on the part of the academia to build such trust once again between various diverse communities in India can offer some degree of social cohesion.

The institutions of higher education in India have been emulating the western modules of learning. They, therefore, could not engage in authentic discourse and dialogue relevant to the needs of the contemporary Indian society. Instead of addressing crucial societal issues higher education has been distancing itself from the common man, and creating an elite class of people. There are very few centres where various disciplines find a common platform for academic interaction. Academicians almost always engage in a kind of self-referral and derive their self image and self esteem from their peers. The usual trend is towards specialization of knowledge systems and disciplines. The teaching-learning process also negatively influences the processes of creativity and innovation as it has considered disciplinary specialities in terms of strong boundaries. Knowledge tilts power towards those who have it, and it is their viewpoint which is expressed. This situation creates a big gap between people at large and those running the show of knowledge industries.

There is an essential tension between the practice of higher education and the goal of social cohesion. The diversification and specialization present a fragmented view of learning. Higher education has alienated people, and there is hardly any scope for dialogue between the higher educational institutions and other segments of society. The system of higher education has increasingly maintained a distance from the social reality. As a result it has now reached a stage that it has become imperative to undertake a critical self introspection and action on the part of all concerned parties.

The problem is partly due to 'technologising' of the educational system as a part of the third world tendency to feature themselves as 'developed' and modern societies. Such haste has met with little protest. Market oriented teaching-learning following the 'western model' has informed the paradigmatic framework pursued by most of the Indian institutions of higher learning. In practice, higher education is engaged in achieving a limited goal of self or personal improvement and has little scope for the pursuance of societal goals. It does not recognize the relationship or interdependence which happens to be an alternative offered by the non-western cultures. The western culture nurtures self contained individualism with fixed boundaries. It views the goal of individuation as the hallmark of maturity. However, this kind of culture has reached a point where progress is turning into pathology. In contrast the Indian view of self is relational or interdependent in which self is defined through one's relationship with others.

Cohesion in society may be pursued in different ways. Pluralistic synergy is one of them. Pluralism assumes that different groups are in harmony and maintain a healthy relationship. Questions of identity are linked with this issue. Societies are comprised of various communities and groups. In any society, therefore, complete homogeneity is neither possible nor desirable. Different groups will always exist and intergroup differentiation is bound to occur. However, the groups may live in a synergy. The diverse groups may allow each other to live with respect, dignity, trust and acceptance.

The western model of individualism does not apply to the Indian setup, and the 'relational' view seems to be more desirable. While the goal of the former is to pursue the development of an egocentric individual or independent and autonomous self this does not hold true for the Indian setting. In the western tradition the pursuit of knowledge was motivated by the desire to subjugate. *Knowledge was power* and therefore opened the potential for control of environment including others. This is in contrast to the view of *knowledge for liberation.* The *pursusharthas* do not deny *arth* and *kam* but do recognize that they should be governed by *Dharma*. They must operate within the frame work of *Dharma*. The notion of *Dharma* provides a context which emphasizes order across all life forms.

The intellectual empirialism and hegemony of the West is still prevalent in various disciplines. Psychology as a discipline offers the best example of Americanization of a modern knowledge system. Other social science disciplines have been somewhat fortunate to engage in the discourses more relevant to the Indian society. The immunity of the pursuit of knowledge to the social reality is largely due to the foreign reference group, emphasis on modernist attitude and positivist outlook towards science. The academicians seek appreciation from disciplinary fellows which reinforces and creates a gap or distance between society and educational practice. Its a truism that the social sciences are conducted by and for the urban educated lot. It does not approach the common man and society at large. A large section of the masses remains isolated and gets only marginal attention. They do not participate in the process of conduct or management of higher education.

Many vital functions of the coming decades including social integration, cultural transmission and transfer of technology heavily depend on the institution of higher education. However, the new global and institutionalized educational system skirts issues such as an ever deepening poverty and the destruction of local culture, presenting through itself a new form of colonization—much more serious in implication than the earlier exploitation by the 'British Raj'.

Top academic functionaries (as well as others top scientists, civil servants etc.) may contribute a credible and sustained leadership in a proactive fashion and look for long term and future generations for whom the present is held in trust. Higher education has to play a leading role in training leaders well equipped to meet the challenges, and are committed to the pursuit of excellence, equity and high ethical standards.

Economy needs educated human resource and, therefore, higher education shares an organic link with economy. There exists a dialectical relationship between the two. It has to be incorporated in the policies governing both economic planning and expansion of higher education with a view to ensure optimum use of human and physical capital.

Higher education is expected to nurture the bonds between self and society. Ideally it prepares for direct self preservation, and to recognize and concede the rights of others, to shoulder obligations, facilitating responsible citizenship and parenthood and for other refinements of societal membership. It is clear that these elements do not constitute separate compartments; they are intricately entangled and inextricably woven to create a responsible and disciplined member of a civilized society. Character building is an essential or core feature of education. Education devoid of character can prove to be disastrous.

Development for a social purpose requires transformation of the society and can be achieved by changing societal values and attitudes. As a nation, India has a great opportunity to become the world leader in integrating science and spirituality. Pluralism or diversity happens to be the greatest asset of India. Understanding and love of diversity and pluralism constitute the first step towards a spiritual age. We must take lead in providing spiritual roots to the technological revolution we see around us. We can take advantage of the great opportunity our tradition, culture and ethos have offered to us. The greatest challenges for converting despair to hope during this century are to be found in areas of Ecology, Economics and Gender equity. A synergetic fusion of political will and action, professional skill and people's participation can help to promote economic growth embedded in the principles of ecology and equity.

12

Relevance of Value Education

The term "value" is often considered as the quality or virtue cherished by a community of people. Values are accepted as ideals to organize and direct personal and social conduct of the people. They also refer to the price or worth of objects and engage with the process of valuing or assessing the quality under consideration. These two meanings of value, however, are interrelated. Value, thus conceived, plays a significant role in organizing the social and mental configuration of society at any given point of time in its history. Today a growing discomfort is clearly visible in the various segments of Indian society that relates to a noticeable degree with, the dissociation between education and values. It has been repeatedly voiced that the linkages between the two are becoming problematic. The worth of formal education is now being seriously questioned. In particular the increasing involvement of school going children in socially and morally undesirable behaviours such as violence, crime, and use of pornography etc. has drawn attention from many quarters of the Indian society. In the last one-decade or so such happenings have evinced a steep rise not only in metropolitan centres, but also in cities and small towns. The villages are also following suit. The social quality of life in general has shown a decline. The development of such trends is an alarm signal as they are posing serious threats to the ideals of equality, equity, and fraternity, which are generally supposed to provide a sense of purpose or meaning to the pursuits in individual and collective lives.

It is a common observation that during the half century long experience of political freedom, India has undergone a large-scale socio-economic transformation. As a country the Indian society has been able to sustain the democratic system of governance and there is a sense of pride and accomplishment on several fronts. These achievements apart, one cannot deny the decline in moral standards, low level of inter group tolerance, violence, disruption in family life, increase in the degree of prejudice and discrimination, and overall deterioration in values. It is held that the changing value concerns are intimately connected with the pattern of social change that has taken place under the impact of westernization and subsequently promoted by the powerful processes of globalization. The practice of education has been either contributory to the prevailing system or has maintained the status quo showing very little resistance. In the course of these developments the value base of education has been seriously hit.

The Current Concern for Values

The resurgence of concern for values as currently witnessed is something that is shared by everybody across all the segments of society including policy makers, teachers, writers, and industrialists. It is happening against the backdrop of an ethos characterized by violence, alienation, corruption, dishonesty, social tension, discrimination, and atrocities on various groups (e.g., women, children, poor minorities, weaker sections). It unambiguously points towards the fact that something has definitely gone wrong. It is considered as an indicator or manifestation of the erosion of values in society. The unhealthy nexus of people with vested interests is ruining the very fabric of society. Such happenings not only interfere with the process of national development but also demolish the faith of the masses and shatter the ideals of those leaders who fought the freedom struggle and nurtured the spirit of democracy. They have an adverse effect on the people at large. Confusion over priorities, value conflicts, and dilemmas are now dominating the everyday discourse in the lives of the people in general.

From Economic Development to Sustainable Development

Impressed by the immediate fruits of industrial revolution, the notions of "development", "modernity" and "science" became central to the process of societal progress in the modern world. The so-called developing countries became engaged in emulating the western models of development, which was largely economic in nature. After the apparent initial success, this mantra lost its grandeur and countries in all parts of the world are now realizing its limitation. The new paradigm emphasizing *human development* and *sustainable development* is being proposed, that takes into account the limits of our planet earth. Today's wisdom lies in working together towards a common future. The limits of economic growth as a viable model for development are now clearly visualized. The increase in violence, ethnic conflict, cut throat competition, and increased environmental hazards point towards a situation that demands urgent intervention in the world view and restoring a value orientation that emphasizes equality, equity, fraternity, and sharing across people and a symbiotic relationship with the environment.

Another related aspect of values, pertains to addressing the concerns of the inner world. The experience of tensions and frustrations and absence of mental peace in individual lives are posing a major health hazard. While the causes for this situation are many, lack of self discipline and realistic vision of life happen to be the immediate causes. The search for values in education needs to address this situation in a positive fashion.

The Contemporary Scenario: Need for a Shared Vision

The development of a self-reliant and strong nations in the twenty-first century demands quality based human resources that may respond constructively to the twin concerns of individual and societal well-being. It is important to recognize that we are simultaneously individuals as well as social beings. Therefore, whether we survive and thrive as individuals or a collectivity depends largely on recognizing the significance of sharing the responsibility as human beings and as citizens of a free country. It has become more valuable in today's world order in which interdependence and complimentarity are becoming key features. The world is now acutely aware of the importance of values that can sustain life on this earth. A holistic vision that allows for a positive expression of universal human spirit characterized by the ideal of *vasudhaiva kutumbkam* is the need of the hour. This age old Indian ideal recognizes the possibility of conceiving the entire world as a cohesive structure of a family. It continues its relevance as a goal for the contemporary world.

India and its Responsibility

As members of a sovereign republic, the Indian people cannot escape from their share of responsibility as humans and as citizens of free India. Let us not forget that the idea of freedom does not mean lack of responsibility. The autonomy made available by the political freedom entails certain duties and responsibilities. Of course, we do enjoy self-determination in the exercise of rights and performing duties. The goals and means to achieve them are to be set by the people concerned. It will be futile to merely follow an agenda fixed by the aliens for their advantage. Viewed against this backdrop political freedom to the Indian nation provided an opportunity as well as posed a challenge to define its priorities and articulate the modalities to achieve them from its own vantage point. It may be recalled that the European model of education introduced in India during the colonial period was meant primarily for sustaining the British Raj and reinforcing its imperial interests.

Quest for Values in the Indian Context

Values are the foundations of social institutions and inform the individual level choices and conduct. Values essentially depict transformation of human nature for better. School as an institution along with family, peer group, and significant others which, provides opportunities to acquire, develop and realize the socially and culturally accepted values. While thinking about values we are confronted with two related questions: (1) who *decides the choice of values?* and (2) *for whom the values are prescribed?* These questions assume great significance in today's scenario. Self-determination by a free and sovereign country is the obvious choice. This was the essence of the idea of *Hind Swaraj* and *Swadeshi movement* launched during the freedom struggle. In fact one of India's brilliant philosophers Shri K.C. Bhattacharya advocated for *swaraj in ideas* emphasizing the notion of learning and thinking on indigenous terms.

Against this backdrop it is relevant to examine the values that have received considerable attention since the ancient period. The indigenous system of education practiced in India had an unambiguous emphasis on the universal values like compassion, non-violence, righteous conduct, forgiveness, tolerance, sacrifice, and benevolence. These values were made part of the practice of teaching-learning in such a manner that a person coming out of that system had to imbibe the values as a part of his life style. The Vedic injunctions required a very high order of commitment on the part of a student.

satyam vad,
dharmam char,
satyanna pramaditavyam,
matri devo bhava,
acharya devo bhava,
atithi devo bhava,
yanyanvadyani karmani tani sevitavayani netarani.

The students were asked to practice and organize their conduct around these values.

It is held that a human being is born with five debts dealing with obligations for all living beings, parents, seers, teacher, and gods. Thus a person was connected with everybody and every being. In fact human existence was conceived in terms of an entity constituted by others. The dynamism and continuity (*rita* and *satya*) provide the framework to understand life. This is very well reflected in the Vedic opennes to ideas that are noble: *ano bhadrah kratavo yantu vishwatah*. This realization resulted in endorsing an intense longing for cooperation, sharing, and living together. The experience of connectivity and effort to maintain it is a significant part of aspiration:

sangachadhwam,
samvadadhwam,
samvo manansi janatam,

samani va akutih,
samana hridayani vah,
samanmastu vo mano yatha vah susahasati

It should be noted that in this scheme of things the place of human beings in the universe is considered highest because they are capable of thinking about others and can organize and connect with them more intensely than any other living being. In other words human beings are great because of the greater degree of responsibility shared by them. It is in this sense that humans are more evolved beings. Thus humans struggle for more than what is of immediate individual concern. Life is viewed as continuity from past and connected with present as well as future. This view empowers human beings for the search of meaning. In this process a person is being defined and redefined in terms of goals that could connect their being with the entire universe. This view inherently entertains a development perspective. It transcends self by well-organized regulatory processes. These concerns go beyond the boundaries of religion, caste and nationality. Developing conscience in a value-involved manner is the foundation of this view.

In the Indian tradition, the discourse on values has revolved around the notion of four broad life goals or purusharthas i.e., dharm, artha, kama, and moksha. Also, most of the religio philosophical perspectives emphasize on observing truth (satya), forgiveness (kshama), non anger (akrodha), patience (dhriti), compassion (daya), self control (dama), non stealing (asteya), limited needs (aparigrah), purity (shouch), control over the sense organs (indriya nigrah), and non violence (ahimsa) in thought and action. Buddhism emphasized on compassion (karuna) and forgiveness (kasama). These attributes employ a relational perspective on human existence. Harmony not only among fellow humans but also across all life forms is the right mode of conduct.

The western scholarship prevailing in most part of the Anglo-Saxon world is characterized by *anthropocentric* worldview and promotes a peculiar notion of a liberal individualism. Such a person happens to be a self-contained individual who is an independent, separate, bounded, and autonomous center of the universe of internal attributes. Competence of such a person lies in being able to influence and control the external environment. Nurturing egocentrism (ahamkar) this *individual* is held as the center of all motives, actions, needs, and desires. This stands in sharp contrast to the Indian view that emphasizes relationship and favours an extended, interdependent, *connected model of human beings* wherein the boundaries between self and not self are permeable. This person is "dividual" and shares its properties with others. It's the sharing and not the uniqueness, which characterizes the Indian notion of personhood. This embedded and connected mode of selfhood seeks self-realization in going beyond or transcending individual or personal concerns.

Learning in the Indian Tradition

The pursuit of knowledge occupies a prime position in the Indian worldview. It is through the acquisition of *vidya* that a person becomes a human being in the true sense. In the absence of transformation through knowledge a human being is nothing but an animal (*vidya vihinah pashuh*). Intellectual culture enhances capacity and stability which enable a person to acquire wealth; wealth so acquired enables to conform to Dharma, which in turn leads to happiness:

vidya dadati vinayam,
vinayaeyati patratam,
patrawad dhanmapnoti,
danaddhamam, tatah sukham.

Thus education is rightly considered as a key to success and well-being. It empowers people by helping to realize the hidden potentials and bringing in excellence in action. Being skilled in action is yoga (*Yogah karmasu kaushalam*). Thus we notice an attitude of respect and dignity for a fine blending of theory and practice in a contextualized fashion.

In its operational mode education has traditionally been conceived of as a process which attends to the task of transforming the whole person in a positive direction. It is expected to enable one and make him competent to pursue and realize all the worthwhile goals and eliminate the negative ones. An educated person rediscovers himself as a self-actualized being. Unlike other resources, which dissipate, the resource of knowledge is renewable and can generate its own programmes for its growth and development. This is why there is so much praise and appreciation of education and of learning-in the scholarly texts as well as with folk literature in India. It is through education or pursuit of learning (sadhana) that one can attain liberation (*sa vidya ya vimuktaye, vidyayamritmashmate*) and realize the true self.

Experiencing the Western Education in India

The salience of education in terms of a value facilitating process as envisioned in the Indian traditions has become a dream. In practice the adoption of western mode of education has proved to be a mixed blessing for the society as a whole. While it did help in certain ways, particularly by modernizing the mind set, it also nurtured a view that created self-doubt and self-denigration as far as the worth of Indian wisdom was concerned. The Indian heritage particularly its scholarship was rendered irrelevant. As a result the process of wholesale import of western knowledge, technology and attitude was initiated. The civilizational contributions by the ancestors from diverse ethnic backgrounds were marginalized. The gap between Indian tradition and contemporary reality largely informed by the Euro-American models has widened. This model, being committed to scientific spirit appeared to be value neutral. In reality, however, it favoured values of *materialism, individuality, control over nature, utilitarianism,* and *anthropocentrism*. Viewing "man as the measure of everything" the idea of an economic man became the guiding spirit.

Schooling in Contemporary India

Schooling or formal education in today's India presents a stark contrast to the ideals noted in the Indian tradition. In the contemporary period schooling has a clear emphasis on skills and information geared towards financial benefits. It does not address the needs and aspirations of the people, society, and nation. It aggravates the problems by alienating people from their own ethos and heritage. The younger generation is becoming more vulnerable. Also, the training of mind in isolation is proving dangerous for holistic development of children. The head, heart, and hand, all three need to be coordinated for the fullest development of human potential. Holistic development requires attention to physical, intellectual, moral and spiritual aspects of development. Development needs to be orchestrated within the framework of values. The holistic development of personality has to address universal values like truth, righteous conduct, peace, love and non-violence. It is now widely held that extra input is necessary to install values in the process of schooling. Identification of the scope of this venture and evolving the modalities for its incorporation in the teaching-learning activities is urgently required. People generally expect that schools and colleges are the potential sites for relevant and effective interventions to arrest the process of moral decay that has set in our social milieu.

Educational Institutions and Value-related Interventions

It is unfortunate to note that a crisis in values is being experienced in a society, which has a very rich and long tradition of learning. In this tradition education itself constituted an important human value. A reflection on the changing ethos of teaching-learning processes in the Indian context suggests that we need to create a sense of humanism, well-being of others, and the nation, and preserve what is good and worthwhile. Success in the efforts towards national development and enhancing international standing of the nation will depend upon a strong value base. A need, therefore, to intervene into the processes of value formation at the gross-root level is often expressed. To this end adding quality to education by bringing values back to the center of teaching-learning process is necessary. Thus, value education becomes a process of restoring the intrinsic worth of educational process in the institutions of learning.

The significance of values in education has been a recurring theme of the reports and recommendations of various

Commissions and Committees set by the Government of India from time to time which review and suggest changes and innovations in the structure and functioning of our educational system. The proposals have been well documented and are full of wisdom. They contain valuable suggestions delineating the necessary changes in the value orientation in the school setting. Their implementation has been, however, very problematic and on account of lack of political will not much has been achieved. Also, the educational ideas and proposals of contemporary Indian thinkers like Mahatma Gandhi, Gurudev Rabindra Nath Tagore, Maharshi Arvind, and Dr. Zakir Hussain could not receive due attention in the course of actual planning of education for India.

While developing values in educational context it should be remembered that adding additional course of study on values will merely increase the workload of students as well as teachers. Perhaps a more appropriate strategy will be to use *values to organize and structure the entire teaching-learning process*. Values should be made an integral aspect of teaching-learning. The lesson in any subject be it mathematics, civics, Hindi, science or English may be designed, deliberated upon, and imparted in such a way that they necessarily *invite value related reflection and action* in a natural fashion. Similarly the delivery of subject content may *have in-built elements* of sharing, participation, and care. By redesigning the modes of interaction between students and students, students and teachers, and teachers and teachers within the school setup, possibilities for developing and reinforcing values may be created. This requires *commitment and determination on the part of teachers and investment of the system to stand by the values they admire*. Schools can and should form a nucleus for positive social change. To this end we need to nurture inter faith communication.

Facilitating Interfaith Communication in Schools

Human diversity is an obvious fact. It is reflected in not only the physical features of body like skin colour, height, eye and hair colour but also in the structure and organization of social institutions, cultural practices, religion, language, and artifacts.

These facets of diversity in the domain of our every day experience have made human life meaningful, vibrant and rich. To a large extent these variations, are human contributions in particular settings. They, however, have influenced life on earth in such a powerful manner that they seem to enjoy their existence independent of human creativity. The variety in religious faiths exemplifies this pattern very well. These faiths are intimately linked with specific climatic conditions, available physical resources, and prospects of economic activities and had deep impact upon the beliefs, rituals, manners, and ways of organizing the entire course of events in the life world. The various faiths have evolved to address a set of needs and exemplify projections of certain basic concerns of life.

Religion has played a diverse role in human history and has shaped the destiny of nations. The presence of diverse religions is the reality of multiculturalism noted in India and many other countries in the developed or the developing world. In general there is aversion towards religion on the ground that it is a force that is disastrous for social harmony and development of society. Religion is often contrasted with science and a strong secular view rejects religion as unscientific and, therefore, non-progressive. The role of religion is being seriously questioned and has become a contentious issue. The secularists assert that religion is a personal matter and should not be brought into public institutions. It deals with other worldly issues such as the soul etc. and is a matter of belief or faith. To them, religion is of little value for enriching human life. The public life should be organized on nonreligious or secular foundations. The main thrust of the secularist position is that the state should not interfere, institutionalize or endorse a religion or be guided by it in treating the citizens. It must remain indifferent to religion. The question of interfaith communication as a facilitator of value education needs to be analyzed dispassionately in the context of social reality and current understanding of the issues.

Recently Professor Bhikhu Parekh has addressed this question and his analysis is very pertinent. Professor Parekh is a political theorist and has been Deputy Chairman and Acting Chairman of the UK Commission for Racial Equality. He has also been Chair of the Commission on the Future of

Multiethnic Britain. In his work entitled *Rethinking Multiculturalism,* (2000) Professor Parekh observes the secularist thesis that citizens should abstract away their religious beliefs and be guided by secular reasons alone is problematic. He remarks:

> "Secular reasons are not politically and culturally neutral. They represent a worldview according to which, crudely speaking, the affairs of this world are autonomous and can and should be separated from those of the 'next'. Since the postulation and separation of the two worlds is a secular construct that sits ill at ease with religious consciousness, and which religious persons do not accept, to require them to be guided by secular reasons alone is to discriminate against them. Secular citizens are able to lead whole integrated lives whereas religious citizens, who are required to bracket out their deepest beliefs, are subjected to moral incoherence and self alienation. This is not to say that they cannot for political purposes suspend their beliefs, but rather that the self-abstraction involved devalues much of what gives meaning to their lives, deprives them of self-imposed moral censorship. Allowing them to be guided by their religious beliefs but banning them from using these to defend their views in public does not improve the situation. It introduces self-alienation at a different level by requiring them to speak in a language different to the one in which they think. And it also subjects them to the disadvantages arising from the need to switch languages and defend their views in what is for them a second language". (Parekh, 2000, p. 323)

Parekh notes that religion remains an important force in people's lives and the kind of secularization that was expected has not occurred even in western societies. Religion, like any other institution or thing, is a kind of experience that has positive as well as negative aspects. While religion has been attributed to many negative consequences there is also a long list of movements and campaigns that have derived their inspiration from religion. Religious people generally seek wholeness in their lives. Religion is often sought after to provide moral motivation and energy. It is also important to note that the faith in the ability of science to solve the mystery of the universe or even to provide solutions to human problems and miseries has declined. Now the scientists are seeking a rapprochement with religion. The nature of science and certitude attached to scientific method has gone down. Parekh notes that the secularist requirement is that citizens should abstract away their religious beliefs. This, however, is not possible on the following grounds.

(1) It will amount to discrimination against religious persons and violates the principle of equal citizenship.
(2) It is undemocratic and ignores the wishes of a large body of citizens.
(3) It is impractical, for there is no way of enforcing it.
(4) It is counterproductive as it is likely to alienate religiously minded citizens from the political system and create a crisis of legitimacy.
(5) It is unwise because it deprives political life of both the valuable insights religion offers and the moral energies it can mobilize for just and worthwhile causes.

Parekh rightly observes that citizenship is not a self-contained area of life with its own distinct values. It is one of several mediums in and through which human beings express and live out their deepest beliefs. Therefore, it cannot exclude what matters most to many of them. Rather than abstract away legitimate differences, political life should find ways of recognizing, respecting, and within limits accommodating them. Parekh thinks that religious education can contribute to the development and growth of a multicultural and pluralistic society. He puts forward the following arguments in support of religious education in schools.

(a) In a society in which a large body of citizens think and argue in religious terms, it is part of the training for common citizenship that future citizens should grow up with some knowledge of their major religious traditions. It will be a valuable tool of inter communal education enabling each group of citizens to appreciate and learn something from other ways of thinking and evolving a richer and expansive concept of public reason.

(b) Religion is a facet of life. The state is concerned that religions should not pose a threat by inciting mutual hatred.

(c) Religion has been a source of many emancipatory movements.

(d) Religion provides a valuable counter weight to the state, nurturing sensibilities and values that are ignored or suppressed by the state.

(e) Religion stresses the quality of inner life and urges human beings to examine the kind of persons they have become.

(f) Along with other social institutions religion helps to sustain the deeper springs of morality.

(g) Religion stresses the unity of the human species and challenges the tendency to limit morality to the territorial boundaries.

(h) It may arouse powerful and at times irrational impulses and can easily destabilize society and create hell on earth. We should not take an unduly rosy or irrationally bleak view of religion. It should be brought to the public realm, made it publicly accountable and subject to democratic discipline.

The teaching of religion in schools can help to enable pupils to appreciate great human achievements. Religious literacy is essential for making sense of the contemporary world. If schools do not teach, the family and religious organizations shall promote the sectarian standpoint. The future citizens must understand how a religious person thinks and reasons. Religion may be taught like other subjects. The approach should be comparative, analytical and respectful. It will be religious education and not indoctrination. Pupils must know how the major religious originated, developed, shaped and were shaped by the cultural climate of the wider society. It should also sensitize them to the similarities and differences between the ways in which different religious come to terms with the contingency of human existence. We need inclusive and religion sensitive secularism.

The learning in school settings has many possibilities and occasions that can be used for interfaith communication. One can creatively utilize the strategies of cooperative learning, role play, brain storming, for this purpose. The content areas and co-curricular activities both can be profitably used. The activities promoting interfaith communication and values can be undertaken through the following:

- Music, drama, dance, painting
- Team work, play, sports activity
- Project work
- Assemblies, campaigns, camps, creating an environment in the school.

The basic concepts behind all the religions are common, only the practices vary. Supreme power or self-consciousness to which man orients itself; divine spark, and the treasure within are equally stressed in all the religions. All the religions emphasize certain qualities or values that need to be inculcated and certain negative aspects of behaviour that need to be controlled or removed. The positive ones include brotherhood, courage, duty, forgiveness, giving, justice, love, obedience, peace, repentance, sincerity, and work. The characteristics to be avoided are anger, evil, hate war, and wrath. Interfaith communication may focus on these aspects of life.

Facilitating interfaith communication will be helpful in promoting dialogue across different communities. This will ensure understanding and appreciation of otherwise unrelated, negative and at times pathological perceptions of different groups. This situation works against the interests of democratic process and spirit. Communication demystifies the ambiguities and nurtures an attitude of mutual respect, and respect for diverse identities. Religious education or literacy also has potential to offer a meta perspective to view the reality of human life. It provides an occasion to reflect upon human condition that is inclusive, universal and humanitarian. Interestingly enough, various religious traditions arrive at some common features of life and maintain certain values that are relevant to all sections of society. They need to be captured and promoted. Schools can play a key role in exploiting this aspect of human capital. They can do a better job by maintaining a balanced approach as their goal is to open dialogue and enrich the perspectives of the pupil.

13

Towards a Holistic Perspective on Women's Education

The need for women's education in the context of India's societal development needs no elaboration. As a nation we are lagging behind other developing nations in the world on the human development index and we urgently need to enhance the potential of women by educating them. It is important for their empowerment as well for laying down a foundation for the future of the society. However, the idealized image and societal practices about women are paradoxical and require serious reflection on our part. In the recent past a number of feminist perspectives have been proposed. They however, often put women against men and argue for adopting a compensatory strategy. Such a strategy tries to redress the problems of women by treating them as competitors. In the long run it may not go very far in improving the state of affairs and overcoming their distress. A holistic view that has space for the feminine as well as the masculine seems to be more viable. The educational planning has to respect the presence of both of these aspects of women's identity. A holistic perspective overcomes the problems arising from the oppositions posited and allows the development of women's potential to a great extent.

Women's Reality: Need for Going beyond the Dualities

Human collectivities are often viewed as dichotomous categories and a conflict perspective is proposed. It is, therefore, not surprising to find comparisons of the two genders and arguments that claim gender discrimination as a major stumbling block in the efforts towards nation-building and societal development. The critique of the prevalent patriarchal system preventing the growth and well being of the womenfolk too is pertinent. The life space of women is peculiarly structured. For instance, marriage with its demand for a new sexual and social life is incomparable to the happenings in men's lives. She has to acquire a new set of cognitive schema for her marital home.

Under the prevailing social order women experience trauma, stress and violence throughout their life in many respects. At the same time they are considered as guardians of virtue and purity so far as morality, chastity and sexuality are concerned. Inequality in terms of gender relations is an obvious fact of the Indian social life. Such inequalities take varied forms in different cultural groups making "gender categories" as heterogeneous and non-static.

At present the incidence of dowry, child marriage, wife battering, sexual abuse, declining sex ratio, and infanticide is on increase. These events are causes of great concern for everyone who wants a respectable place for women. The adverse consequences of these happenings necessitate a move to reconstitute the discourse pertaining to women, family and motherhood and make a case to fight for the cause of women from a broader perspective. This should be taken up on a priority basis as it involves the status of the entire society rather than women only.

Images from the Tradition

In the Indian tradition, community or national identity is considered practically inseparable from the identity of women. Motherhood is taken to symbolize the entire community. The Vedic declaration that earth is mother and we are its children (*Mata bhumih putroham prithivyah*) and Vedic women like Gargi and Maitreyi who participated in disciplinary debates (*Shastrartha*) do indicate the presence and participation of women on equal footing. Bankim Chandra's *Vande Mataram* clearly expresses the traditional Indian concerns for mother. Any assault on mother/women is seen as an assault on the nation/community.

The value of women in an agrarian society organized by a predominantly collectivist ethos was rightly expressed in the saying that "gods live where women are respected" (*Yatra naryatu pujyante ramante tatra devatah*). Women are considered as the prosperity of the house. It was declared that women should always be protected (*Yoshito hi sada rakshyah* – Mahabharat Adiparva 110/112). Gods have been presented in feminist ways. An autonomous female principle *Prakriti* or *Shakti* has been posited. Female divinity such as *Durga, Lakshmi, Saraswati, Kali, Sita, Ganga* and a host of others are worshipped by (even modern) Hindus, irrespective of their political commitments to communism or socialism.

However, due to various historical, cultural and economic reasons, the real life situation in which women live has deteriorated. As mentioned earlier their treatment during different stages of life is often full of discrimination and exploitation. The result is that a large section of women are now subjected to deprivation and exploitation on various counts. They are treated as dependent and have limited scope for self-enhancement. Their access to and control of material as well as psychological resources is very limited. They suffer a lot in various roles throughout the life course.

The Discourse on Women: Changing Contours

If we examine the discourse on Indian women in modern times, beginning from the period of social reform, nationalist period to contemporary times we notice a number of contradictions. These contradictions essentially reflect the problems pertaining to the process of modernity that has been male centered. As an alternative, the feminist view is often proposed and advanced, which is rooted in a reactionary mode of thinking. In such efforts the emphasis is more on engaging in endeavours implying compensatory mechanisms to address the disparities. Such a short-term perspective strengthens the tendencies of individualism, consumerism and the neglect of ecology. It does not take into account the social system in its entirety. It also ignores the experiences and realities of Western societies where mental health problems, marital discard, familial dysfunction, and social pathologies are on the increase.

Need for an Inclusive Perspective

Against this backdrop it becomes necessary to plan education for Indian women that may sustain individuals as well as the social structure. It should empower women to undertake the responsibilities as an integral part of society. They should be neither victims nor exploiters. A joint and meaningful engagement with the processes of social life is warranted. Such an approach makes family and social lives viable, enduring, and capable of resilience. Only they can sustain the social and developmental processes. A holistic approach involves a balanced integration of femininity and masculinity.

A more fulfilling and humane education for women therefore shall address not only the feminist concerns but go beyond it because it does not posit the essential dichotomy between women and men. It has to realize the incompleteness of each of them as isolated entities and must overcome the conflict and tension between the two. Such a view does not deny the existence of discrimination and atrocities to which the women are subjected to, in many communities. It, however, does promise to bridge the gap between the two because it is only the relational fabric at the levels of individual, family, group, organization and nation that can ensure a positive view of humanity.

Concluding Comments

Education should be made a process that facilitates the goal of pursuing individuality as well as relatedness. It should try to establish an order amidst the crises of the contemporary period. Fortunately Indian culture does have potential and promise. Its moral systems do have encompassing schemes which have in its fold the whole range of values. The system of *Dharma, artha, kama* and *moksha* is a grand abstraction. It integrates almost all life goals as well as prescribes the modalities to organize our conduct. We need to come forward and bring these values to the centre stage.

The holistic perspective recognizes the complementarity, interconnectedness and interdependence among the different subsystems. It also respects the role of embeddedness of existence in the eco-cultural context. Negotiating the balance between tradition and modernity and relationality as well as individuality is the need of the hour.

At present, women's education closely corresponds with the pattern of social class variations in society. It is still inaccessible to a large section of the population. The problem, however, is more serious because the kind of educational system that has been visualized for women is not as inclusive in character as we require. With a view to promote women's participation in the mainstream and nurturing an integral personality we need to rethink and align the processes of education in a context friendly, responsive and inclusive manner.

14

Socio-Psychological Perspective on Caste Related Prejudice*

In view of the prevailing notion of prejudice there is hardly any society which can be said to be free from prejudices. Prejudices are largely related to social categorization, a fact characterizing most of the societies and communities. The social structure may be stratified adopting single or multiple features such as religion, economy, race, and power. Such classifications are as inconvenient as they are intractable (Nandy & Kakar, 1980). In a general sense prejudices are conceived as prejudgments regarding persons, objects or ideas based on incomplete and inadequately substantiated data. Prejudgment helps an individual to simplify the judgment process as well as the social world. In day-to-day social interaction people frequently act on the basis of attitudes which they hold about groups, and generalize to all the members of the target group. It is in such cases that we refer to ethnic prejudices which are used mainly in the negative or pejorative sense. Such stereotypes are pervasively found in cultures and universal status of prejudices has been envisaged. However, cross cultural research does not show universal patterns. As a result attention is paid to the cultural context (Jones, 1990). Prejudices in the form of stereotypes are defined as normative conceptions about in-group and out-group members (Pepitone, 1986). Members of a given culture share these stereotypes and adopt them.

---

* The ideas were developed in collaboration with Prof. Uday Jain.

The Changing Scenario of Caste

Caste or Jati is a very strong social institution constituting Hindu society. It is linked to the *social order which* was envisaged to pursue life goals (*purusharthas*) was organized in a hierarchical structure of *Varnas* (Social Groups) namely Brahman, Kshatriya, *Vaishya and Shudra*. These groups were supposed to perform specialized duties and engage in specialized occupations. Such a stratification reflected social segmentation based on occupational categories. At the same time it was also viewed as a proper *social order for the growth* of individual and society. The *Varna* system was gradually linked to birth and provided the ground for the caste system (*Jati*) which came to be practised as hereditary, hierarchical and endogamous.

In contemporary life we find that some of the features of caste system have changed. For instance, the linkage of caste with occupation has weakened, economic status of the downtrodden has improved in some sections, dining in restaurants and hotels in urban areas etc. has become open to all castes. All this has facilitated inter caste interaction. However, Hindu high caste people still show resistance to change particularly in the marriage system. Majority of these people are continuing their lives as members of different caste groups in which they are born (Kuppuswamy, 1975). Moreover, the efforts of the Government to extend the welfare provisions to the scheduled castes, scheduled tribes and backward castes have met with violent resistance from the upper caste groups. The traditional caste system in India has been interacting with the democratic-political system of the country and the outcome of this interaction reflects hypocrisy on the part of high caste elites.

The origin and development of caste system in India is explained in many ways. The racial theory (Risley, 1969) maintains that the Aryan race tried to maintain purity against the idolatrous faith of the aboriginal tribes which gave rise to the caste system. Such a view of racial purity tended to deepen the traditional prejudices of caste among Hindus regarding their intrinsic superiority. However, as evidence suggests all people have mixed pedigrees. The hypothesis of racial or genetic differences is untenable. The occupational theory suggests that castes, originated in the course of the division of labour or specialization of various functions in the social order. This view, however does not explain the complexity of caste. For instance, large number of crafts are much later in their origin than agriculture which has been held in high esteem as compared to all other occupations. Dumont (1970) has emphasized the importance of ideology in ordering the social system with all its inegalitarian social relations. In the case of caste system the ideology of hierarchical social organization reinforced by the religious practices and rituals was very crucial. In Hindu society, primacy is given to the whole, and this whole is hierarchically structured (Sinha, 1982).

Thus it is obvious that none of the above mentioned theories receives unequivocal support from empirical studies and logical analysis. Singh (1988) has summarized these views as cultural and universal and has emphasized the structural and particularistic treatment of caste. Thus caste is treated as an institutionalized system in the society which provides hierarchically ranked hereditary groups for marriage, occupation, enforcement of cultural norms and values. Due to various socio-historical reasons the number of sub-castes and castes has increased manifold. However, the myth of the 'varna' and its hierarchical order has persisted through centuries. The institution of caste existed over a much longer period because it was more adaptable. The new features were accommodated in this system without compelling the system to lose its characteristics.

Psychological Perspective on Caste Prejudices

Social psychologists have included caste as a relevant variable in studies on a variety of issues including deprivation (Sinha, Tripathi, & Misra, 1982), socialization (Anandlakshmi & Bajaj, 1981), interpersonal attraction (Prasad & Prasad, 1980), cognitive abilities (Rath, Dash & Dash, 1979), conservatism (Sirkar & Hassan, 1973) and academic achievement (Singh 1980). These studies generally tend to suggest a significant positive relationship between caste status and competence in functioning in various areas.

The study of caste prejudice was initiated by Rath and Dash (1957) who reported more progressive and liberal attitudes towards caste system in higher caste people (e.g., Brahmins) than in lower castes. Moreover, high caste people showed lesser negative attitude towards untouchability than scheduled castes themselves (Natraj, 1962; Rath & Das, 1957). The in-group-out-group differences in the case of caste are interestingly very similar to those reported in the western studies on prejudice. Thus lower caste people possess a negative self image in comparison to high caste people (Rath & Sirkar, 1960). Despite similar changes in these stereotypes higher castes usually maintain age old prejudices against lower castes (Anant & Premshankar, 1966; Sinha & Sinha 1967). Such stereotypes determine the preference for association with the members of different caste groups and social distance (Sinha, 1971).

An extensive survey of beliefs and options regarding caste in college students conducted in Pune by Paranjpe (1970) revealed lack of belief in the doctrine of *'karma'* and 'rebirth' among majority of the students. *However, the extent of social* distance was greatest between *Brahmans* and *Harijans.*

The attitude of high caste Hindus is reported to be more modernized than their low caste counterparts (Hassan, 1976). At the same time other studies (Prasad, 1979; Sirkar & Hassan, 1973) have reported greater prejudice in high than low caste groups. A few other studies have found no difference between high and low castes in the intensity of caste prejudices (Raj. 1980). Discrimination towards low caste people has also been reported by Prasad and Prasad (1980), where punishment for hypothetical criminals varying in caste was to be assigned. The high caste subjects did not suggest differential treatment for the high and middle caste criminals. (Singh, 1988)

While these studies focus on psychological factors they tell very little about the role of contextual factors. Psychologists in India (e.g., Sharma & Anandlakshmi, 1981) have examined socialization of caste attitudes. Referring to American and European studies these authors have highlighted the role of parents and supported the ideas of Allport (1954) concerning social learning of attitudes.

It has been noted that the awareness of caste in children emerges around 4 to 5 years (Singh, Singh & Singh, 1960; Vyas,

1973) which stabilizes around 7 years of age. The role of parents and particularly of mothers has emerged a significant factor in the development of caste prejudices (Hassan, 1983, 1976; Sharma & Anandlakshmi, 1982). Caste awareness includes identification of castes, discrimination by caste and finally categorization of people into different caste groups. Children undergo the actual experience of behaving discriminatively on the basis of caste groups under the direction of parents and other elders in the family. Empirical evidence shows that moderately severe punishment was enforced by parents for children's inappropriate behaviours toward various caste groups. In this way the cycle continues. Other discriminatory behaviours of children include acceptance of food, choice of friends' home visits and participation in religious and social activities etc. In other words caste awareness is present in children and influences social participation in significant ways (Tiwari & Misra, 1980). In general, children show more liking and greater interaction with their own caste children than with those from other caste groups. These findings reveal affective and behavioural manifestations of prejudice.

Indian children are exposed to discrimination directed towards lower caste groups from very early age. More significantly, the children of Hindu families learn the social values embedded in the day-to-day life. They also learn discriminatory behaviour expected from them (Sharma & Anandlakshmi, 1982). The hierarchical social structure dictates interpersonal relationship based on their respective caste positions. Also caste and economic class, in majority of cases often overlap. The lowest economic class is largely constituted by the lower caste groups.

Psycho-social Roots of Caste Prejudices

A question is raised that why could democracy and industrialization not bring any substantive change in the occurrence of caste prejudice? An answer to this question cannot be sought in terms of various universal theories e.g., symbolic racism, contact hypothesis, social identity theory (Tajfel, 1997) etc. It is not the effect of simple social learning during childhood that causes prejudice. It is much more than that. Probably the answer comes from the cultural meaning system based on religio-philosophical knowledge and practices which have continued for thousand of years. Caste prejudice can be comprehended within the framework of a conceptual system based on the beliefs in the principles of *karma, varna* and *dharma*. These interrelated concepts provide a social order based on birth due to past *karma* in particular caste and duties according to the hierarchical status of caste. Thus caste provides a social identity to the person. This identity is supported not only by a person's caste alone but also by the religious sanctions. Members of various lower caste groups not only belong to low caste groups but also have to learn to behave submissively with upper caste groups. Social identity seems to be reinforced by the norms related to caste. Evidence to this effect can be found in the studies on attribution where ability attributions varied with the caste of children (Sharma & Tripathi, 1988) and the studies showing vast majority of people in political and social life do not consider caste as an evil (Srinivas, 1962).

The stereotypes regarding *Kshatriya (Rajputs), Brahman, Vaishya, Shudras and Harijans* prevail (Anant & Prem Shankar, 1966) inspite of education, industrialization and other democratic institutions. The changes in caste system typically enhance differentiation along caste lines. In reality caste organizations and subcaste formations are on increase. Effect of modernization on caste system can be explained in terms of contextualization (Ramanujam, 1990; Roland, 1988). Thus, one can behave as a western man in one context and as an orthodox traditional Indian in another context without experiencing any kind of dissonance. Applying this analysis to caste behaviour one can easily notice that Indians advocate democratic values (e.g, equality, fraternity, liberty) and practice aspects of caste system, where, equality and fraternity are practiced only within own caste groups. The low caste groups are considered as responsible for their present adversities.

Even the structure of extended family is hierarchical and each family tries to enhance its reputation within its own caste (Roland, 1988). Belief in hierarchy is inculcated in children from the very beginning. The developed self of an Indian adult is not 'I' but 'we' which is multifaceted and relational (Ronald, 1988). This kind of self has an identity where the individual is embedded in collectivities (Tripathi, 1988). In spite of caste prejudices, social and religious functions are performed with the cooperation of people from different castes.

Recent analysis maintains that stereotypes are normative conceptions (Pepitone, 1986). Thus, caste categories constitute an objective reality in the sense that they are adopted by individuals and form a salient part of people's cognitive structure. With the presence of this objective reality (as cognitive schema) the caste prejudices and discriminations seem obvious and inevitable outcomes. This line of thinking is highly individualistic and cannot help solving the problem. The culture specific analysis of prejudices demands explorations into the belief and value structures shared by the different groups.

Caste prejudice is a complex phenomenon, the amelioration of which is plagued by deep rooted conflicts. At one level standing by one's own caste obligations and observing various practices still continues to be a part of normative social behaviour. At the other level, in view of modern aspirations of a castless society, the same becomes an evidence of prejudice and discrimination. Despite westernization and industrialization the caste system is continuing and manifesting itself in the form of different realignments in the political and social arena. Whether caste linked behaviours should be treated as prejudiced or not, depends on the forestructure of people's religio-social beliefs. For instance, the social distance maintained across caste groups renders certain behaviours acceptable. The same however, lies beyond the range of normal social behaviour, once we accept the state of casteless society as desirable. The locus of prejudice is not in the individual but in the social and cultural practices. In other words, prejudice is a contextual and systemic phenomenon. Interventions for the removal prejudice should, therefore, focus on systemic processes rather than modification in the individual level cognition and/or behaviour.

15

Social Problems, Psychology and Social Policy*

In recent years social sciences have evinced a keen interest in contributing to the solution of social problems. These problems change with time and are intricately related to the social context. However, under the impact of the natural science paradigm, psychologists have maintained a value neutral posture and have remained preoccupied with the analysis of variables and their interrelationships in a decontextualized manner. Consequently, socio-historical and cultural factors that shape social reality have largely been ignored. This is perhaps one of the most important reasons why the social sciences have not been very successful in dealing with social problems. In some situations researchers have merely maintained the status quo and even confounded the issues.

Viewing social problems as situated in the socio-historical context increases the probability of designing and evolving successful solutions and necessitates that social problems are to be defined and solved anew in every generation. This is not an easy task and may not even be considered desirable within the prevailing natural scientistic framework. In this context, the task of social scientists is to offer different perspectives on these problems and suggest diverse options. Several drastic changes need to be incorporated in the discipline in order to enhance the understanding of social problems and to transform the prevailing diagnosis. This constitutes the first step in the

* The ideas developed in collaboration with Dr. Suneet Varma.

direction of significant participation in the arena of policy making. The final outcome would be increased social action and empowerment of the disadvantaged sections of society. Against this backdrop, let us examine the nature of social problems and policy, and evaluate the psychology-policy interface.

Social Problems

Social problems are rooted in the socio-cultural and historical conditions. Therefore, they have to be defined in different ways during different time periods. During most of the twentieth century, social scientists have largely ignored social history in their academic discourse and held the view that like other sciences, problems once solved become obsolete. There is now increasing realization amongst social scientists that social problems, by nature, are such that they need to be solved continuously and in a different way each time (Seidman & Rappaport, 1986). Thus a proper understanding of contemporaneous values becomes very crucial, for it provides, the ground for formulating problem-solving strategies that are effective in the prevailing context.

There are a number of ways in which social problems can be characterized and identified. Henley (1986) has given a useful framework outlining the following characteristics of social problems:

(1) These problems affect a large number of people.
(2) Some objective condition exists that can be pointed to.
(3) Some people are aware of the objective condition and are concerned about it.

Any condition which shares these characteristics indicates the existence of a potential social problem. In addition, Henley (1986) suggests that there are four major conceptual approaches for recognizing social problems, viz., functional, normative, objective and subjective. The functional approach assumes that there is a state of balance existing in the society and conditions which disrupt this balance are identified as social problems. The normative approach is based on violation of the norms prevailing in the mainstream society. Here an attempt is made to locate those individuals whose norms have been offended and to find out in what way these individuals have been effected. The social problems exist for those individuals whose norms have been violated. The objective approach recognizes the existence of social problems when empirical evidence points to certain intolerable conditions which have come about cumulatively. Over a period of time the conditions worsen to such an extent that it cannot be ignored any longer. The objective approach by itself is inadequate for it does not take into account the perception of individuals. Unless a situation is perceived by a significant number of persons as a social problem, for all practical purposes, no problem exists. This makes it necessary to analyze a situation in a subjective manner, in terms of the perception and cognition of the social issues involved, by the concerned people.

Nature of Policy

Policy formulation essentially comprises of charting a definite course of action to guide and determine present and future decisions. A policy, therefore, demands inputs related to decision making for group(s) of people. A policy is related to certain social problems and issues which may have long-term or short-term implications with backward and forward linkages. The policy makers are entrusted with the task of managing the affairs. A policy aims at modifying a situation towards a more desirable state of affairs. All policies have their origins in problematic situations which are not tackled within the system itself. When the self regulatory mechanism of a system fails the governmental agencies come in to correct the situation through appropriate measures. Thus policy is a response to the perception of problems by the government and hence such decisions are primarily not academic or theoretical.

In general, policy-decisions are related to multisystemic considerations. Social, cultural, economic and political considerations usually co-determine the policy decisions. The degree of relevance that an academic discipline has towards policy corresponds with the degree of openness of a given system or subsystem in relation to other systems. As such, a policy is a statement of a set of prescriptions and how a discipline relates itself to policy is primarily determined by the goals of the policy itself. For any discipline to contribute towards policy making, its objectives must be congruent not only with the social and political ideology but with the aspirations and requirements of the target group as well. This gives the social sciences, including psychology, only a limited role in the formulation of policy. The factors which confound the role of social scientists in policy making include certain biases inherent in the disciplinary practices of social science : We are concerned here with the discipline of psychology and therefore, our discussion will largely address to this field.

Errors of Conceptualization

Most of the research and theorizing in the social sciences tends to perpetuate the status quo, the main reason of which lies in conceptualization errors. As Seidman (1978) has indicated, these errors are: *societal values, officialdom, and nature of social science.*

Societal Values—The error of societal values refers to the conformity of social scientists to the existing power structure (administrative-political). Social scientists usually accept the power structure's conception of what is right and what is wrong in the society and these values are then imposed in their research, resulting in a self-defeating process. For example, the western system of education has been adopted by the government in this country and along with this have come many of the concepts and methods which are alien to our cultural milieu.

Officialdom—The error of officialdom refers to the reality that key members of the administrative-political system have a vested interest in holding on to their jobs and the power and status that go along with it. Therefore, they are not likely to disturb the status quo and come up with permanent solutions which in the long run could cost them their jobs. This is only natural because if fool proof solutions are found, the services of such officials would no longer be required. In short, social scientists first accept the power structure's norms about what is right and wrong in the society and subsequently when solutions are proposed to the officials, they tend to accept only non-threatening approaches. These two factors seriously limit both the choice and method of dealing with social problems.

Nature of Social Science—The conceptualization error of social sciences extends the error of officialdom to the domain of social scientists who, along with key members of the power structure, have a vested interest in retaining their jobs and means of livelihood. An example from clinical psychology makes this issue clear. To date, it has not been firmly established that psychotherapy actually works. Mainstream clinical psychologists maintain that psychotherapy is effective but research studies indicate otherwise (Leahey, 1991). A large number of studies have concluded that psychotherapy under a clinical psychologist seems to be no more advantageous than self help (Priolieu, Murdock & Brody, 1983; Zilbergeld, 1983). Studies have also indicated that psychotherapy may not be safe because most clinical psychologists are not competent and there have been many cases where the patient's condition has worsened due to therapy (Zilbergeld, 1983). Gergen (1990) concludes that "there is reason to believe that in the very efforts to furnish effective means of alleviating human suffering, there are respects in which mental health professions simultaneously generate a network of entanglements for the culture at large. Such entanglements are not only self-serving for the professions, but add exponentially to the existing sense of deficit" (p. 254). In addition, social scientists as part of the larger socio-political system, are bound to it in certain essential ways. In this context the following factors need attention.

Funding Agencies

Funding of most research in social sciences is done by government bodies. Social scientists are thus obligated to such bodies and have to conform to certain terms and conditions and priorities. For example, definition and choice of research problems must be formulated using terms and concepts which the funding body is familiar with and which are acceptable to them. In addition, only official data must be utilized. As we already know, government agencies have a vested interest in adopting appropriate definitions and data.

Constraints of Academic Administration

Apart from constraints imposed by the funding agency, academic administrations also have certain policies which must be adhered to by the social scientists. In general, the administrators desire that their faculty attract maximum fame with minimum controversies. This, of course, requires supporting the status quo. As Moore (1973) has observed the social scientist is hampered by the pressures of the university administration to publish, procure maximal funding and "avoid generating controversies that would offend the non-university public" (p. 69).

Paradigm Constraints

There are certain intra-disciplinary features which make the efforts of psychologists less relevant to policy formulation. The paradigm constraints come under this category. In general, the paradigm of psychological research envisages uni-directional causation. It uncritically emulates the model of physical sciences. As Maruyama (1974) has indicated, we assume that like physical sciences our universe consists of material substances. These substances persist in time, they obey the law of identity and the substances are classifiable into mutually exclusive categories. Categories may be divided into sub-categories and may be combined into super-categories. The categories are independent from things and higher than things. The methodology with synchronic emphasis and focus on entity rather than system has drastically reduced the effectiveness of psychological research.

The diachronic theories appear to be more useful. There is a need of methodological and theoretical pluralism. Psychology has been following the model of parametric sciences which assume that the properties referred to by the variables that describe the system vary sharply while retaining their identity. It is also assumed that parameters are such that one can hold constant any one parameter and vary others and the property represented by that parameter will remain unaffected by the observation. In contrast, structural sciences constitute internally related structures. In a structured entity, the component parts derive their meaning from the other details to which they are internally related. The familiar paradigm of psychological experiments follow the parametric sciences. The development of ethogenic movement and ethnomethodology have made some innovations. However, in general, the research process neglects the system, isolates the variables and examines the relations among abstract variables.

The biases prevailing in various forms across different phases of the research process further delimit the usefulness of our research efforts. The work for policy demands a dynamic and sometimes paradoxical research strategy. Since policy makers have to view a given problem in a wider context and have to think in terms of short-term, medium-term and long-term consequences, their urgency and structural perspectives become incompatible with the narrow perspective and synchronic paradigm of social sciences which often emphasizes a selected or chosen part of the system in isolation. This does not mean that we are not interested in the systemic perspective. Rather, it is an inevitable consequence of the methodological and theoretical constraints which demand treatment of structural problems from the parametric perspective. The parametrization of structural variables has been accentuated by the research design and static analysis. Almost all the variables in psychological research are systematically linked and do not fulfil the requirements of parametric variables.

In summary, social scientists tend to maintain the status quo (socio-political-administrative) due to their vested interests in holding their jobs, constraints of funding, university administration pressures and paradigmatic constraints. These problems seriously jeopardize any sincere effort towards the formulation and subsequent solution of social problems.

Psychology-Policy Interface: A Critique

The picture that emerges from the present analysis is rather pessimistic regarding the role of psychology in policy formulation. Nevertheless, psychology can play a significant role in contributing towards areas such as teaching-learning, child-care, management, communication, etc. These are areas in which psychologists can contribute substantially to policy formulation. In the present transitional era, ambiguities prevail in these fields and various sections of society are looking towards psychologists and other social scientists for guidance and help. However, some major intra-disciplinary changes are required before psychology can play an active role in the realm of social problems and policy making.

As Singh (1982) has indicated, a policy science is required to (i) incorporate normative goals of development into specific objectives of research which may contribute directly to policy formulation, (ii) generate data on a periodic basis for the purposes of planning, implementing and monitoring the progress of development schemes, (iii) conduct follow up research, (iv) examine the value loads of social policy, and (v) build up and maintain a professional body to regulate its interfaces with other policy sciences, policy makers and change agents. A close examination of these prerequisites of a policy science reveals that psychology does not meet these criteria to an appreciable degree (J.B.P. Sinha, 1988).

A careful scrutiny of psychological research of the past several decades reveals that our research efforts have been primarily non-policy oriented. They are mostly conceptual and theoretical and have examined the relationship between abstract variables. Most of them have their origins in the context of some theory and not in the real life context. Like true scientists we have been interested in theoretical relationships and not in problem solving or decision-making. Of course, our interest has gradually been shifting to the processes and theories which are in the direction of approximating social reality. This is due to our increased concern for overcoming the threat to external validity. If policy has to assume a high priority goal for research, it would require corresponding changes in the configuration of the discipline.

Till recently research endeavours in psychology have remained either unconcerned with the social problems or have not developed the mechanisms for maintaining a meaningful and continuous dialogue with society. We have used society for our scientific activity and have bothered very little about the possibilities of using our science for society. The outcomes of research remain largely inaccessible to the society or policy makers and are circulated only among the colleagues within the discipline. Moreover, our best research studies, those which are usually published in foreign journals, remain inaccessible to a large section of the members of the same discipline within the country. So far, no conscious attempt has been made to go to the people or masses with the fund of knowledge which has been generated through research.

The dissatisfaction with research is being expressed at various forums. However, there is little effort to address our society as well as the governmental policies. This is happening despite the fact that we enjoy considerable freedom and autonomy in selecting the research problems. Apart from sponsored research funded by some external agency there is very little binding on the choice of problems, selection of method, publication and dissemination of research findings. This implicates that the research culture prevailing in psychology is designed by the academicians and change in any positive direction has to come from within the academic community.

Theoretical research strives at formulating nomothetic laws concerning the relationship among variables, while applied research focuses on prediction and effects. Applied research often provides basic information for decision making while theoretical research deals with the analysis of abstract concepts. Unfortunately, applied research in India is also more biased in favour of theorization. In recent years, increasingly greater disillusionment among psychologists is noted with respect to psychology's response to policy demands. A close scrutiny of the field reveals a number of factors within and outside the discipline which are responsible for this sorry state of affairs. The colonial background, borrowing of concepts, tools, methods and problems from Euro-American psychological traditions, lack of inter-disciplinary dialogue, inaccessibility to indigenous traditions of knowledge and insensitivity to the crucial aspects of social reality are some such factors.

In this context, Naidu, Sethumadhavan and Pandey (1988) have questioned the status of psychology, both in western as well as Indian context, as a significant social science which can contribute towards policy making at the national level. They point out that there is no agreement amongst psychologists about the basic questions which the discipline addresses and attempts to answer. They have argued that "psychology, especially in India, has been an elitist science which has ignored issues such as poverty which concern the vast majority of our people. The motivation of Indian psychologists is rooted in career-advancement and establishment of psychology as a science, while the inappropriateness of the scientific paradigm for psychology has largely been ignored. Thus the discipline badly needs to develop self-oriented methods and to loosen the hold of pseudo-scienticism, blind adherence to the scientific paradigm. Psychologists need to have greater knowledge about themselves and what the discipline has to offer, before demanding to be consulted in formulation of national policy. Psychologists in our country have lost touch with the Indian reality and are unable to emphathize with their fellow countrymen. Before psychology can begin to contribute to national policy making, it needs to develop an internal policy which will define and guide the discipline. Such a policy must lead to the development of paradigms other than the orthodox scientific paradigm."

Enough has been said about the shortcomings of psychology in the domain of social problem solving and policy formulation. A self-critical attitude and an open mind is the first step towards a positive direction. In the following section, some changes are proposed which can be incorporated in the discipline to improve its potential to contribute to social policy formulation.

Towards a Better Future: Some Suggestions

Social problem solving provides opportunities for psychologists not only for testing their theories and understanding, but also contributes to the ways and means of reducing human suffering. This, however, demands extensions and revisions of the role of a pure, value detached and objective scientist. The social world is certainly not a scientific world.

The strict adherence to empiricist epistemology has led us to adopt a strategy under which we ask questions that suit our preferred methods. In addition, we often miss the point that applying scientific knowledge in the social field is both an art as well as a science.

Each society defines and tries to cope with its social problems differently and this changes with time. Psychology, due to its identification with 'science', has derecognized the significance of social history. Consequently, the gap between science and human affairs was also not recognized. For instance, choices in the social arena are not based on data. Indeed, they are determined by tradition, values and world-views held by the people. This implies that making policy relevant to the needs of society and solving social problems would require broadening of our perspective with the help of inputs from allied disciplines and greater sensitivity to the context which often appears extraneous but produces systematic variance in the pattern of behaviour.

The preceding discussion serves to illustrate some of the important aspects of the reality of psychological practice which have a bearing on social issues and policy making. Psychologists have a dual role to play in the domain of national policies. First, they can provide the policy makers with a dynamic analysis of the psyche of the people rooted in Indian culture. Such an analysis has been impeded by a blind adherence to the scientistic paradigm. Alternative approaches such as social constructionism, hermeneutics, indigenous psychology are much more appropriate for this purpose. It is only when an authentic picture of the Indian social reality is available that appropriate and effective policies can be formulated. Again, psychologists can play a significant role in the implementation of policies by the application of appropriate behavioural techniques. At present, there are a variety of psychological tools and techniques available which can be used for behaviour modification. Effective application of such methods for behavioural change presupposes an adequate understanding of the dynamics of individual and social life.

Some examples will make the above mentioned issues more clear. Family planing measures have largely failed in our aspect of the reality of the target population, i.e., the fact that the target group does not believe that it is in their interest to have less children. Any attempt towards family planning must address this core issue. First, the concerned individuals must be convinced that it is better to have less children and only then will they be inclined to adopt measures to that end. Psychologists can play an important role here in bringing about attitude change via effective communication. The issue of increasing national productivity is another significant area. The trend has been to explain low productivity as a consequence of low achievement motivation amongst Indians. Subsequent efforts to exploit achievement motive towards national development have met with little success. The reality is that different cultures hold different meanings of achievement. What is valued in one culture may not be valued in another. In the Indian context, a relational view of the individual and family concern is given more importance than personal achievement. Therefore if we wish to increase productivity, efforts must be made to tap and exploit this dimension of the Indian reality. One of the reasons why Japanese management is so effective is the recognition on their part of the importance of family life and group goals for the common man of Japan. A Japanese will seldom work hard for more money but when it is a question of what is good for the family or family honour, the story is quite different.

The last decade of the twentieth century marked a turning point for the discipline of psychology. We are beginning to come to terms with the crisis which the discipline has been undergoing but sustained and critical introspection is required on the part of Indian psychologists. Before we can contribute towards national policy making, we need to be clear about our motives and goals. The fundamental issue for psychology is that of developing a paradigm which is appropriate for our purposes. The future of any discipline depends on its students. As Kuhn (1970) has rightly pointed out, it is the young scientists who take the significant step towards a new vision of reality. The future of psychology rests on the shoulders of today's students. We need to be more honest with the serious students of psychology and share with them the limitations and constraints of the orthodox scientific paradigm. At the same time, we must expose our students to the alternative approaches which are emerging and encourage them to think along new lines which would be more productive in the applied setting including the domain of social problems and policy formulation.

16

Science, Culture and Subjectivity
Relocating the Discourse on Psychology

The idea of "science" and its application to psychology is an interesting illustration of the practice of modernist worldview in the world of academia. Working within this worldview the spirit of science tries to locate knowledge outside or apart from that which is to be known and from the activity of knowing. The power of participants in this kind of knowledge generation is drastically curtailed and marginalized. But certainly there is more to the truth than what the scientific method can offer. However, the claims of universality of scientific knowledge and its independence from the context appear vacuous when one finds that acceptance of some form of pre understanding or theoretical fore structure is an essential ingredient in any kind of scientific mode of knowledge generation. Also, scientific communities are located in specific historical settings, which decide the legitimacy of knowledge claims by determining the relevant criteria. All this renders the alleged methodological certainty and universality in the scientific enterprise questionable. The idea of natural sciences as "supra historic", "neutral enterprise" and "the sole mode of acquiring true knowledge" appears problematic. However, by invoking transcendence of history and culture, the proponents of such a view argue in favour of scientific knowledge as transcendental. This serves the goal of achieving mastery. In essence it has become a part of modernist philosophical discourse, which constitutes the self-understanding of science and its inflated claims. The main voice of psychology exemplifies this state of affairs.

The Project of Scientific Psychology

The construction of psychology as a scientific discourse has been the major factor informing the discipline's self-understanding. As a result psychology became a science dedicated to the discovery of behavioral laws. Psychologists in the mainstream have been vigorously pursuing the goal of achieving regulation and control of behavior of *others*. This became the core concern of psychologists and constituted the mode of conducting psychology. This was in line with the position that knowledge consists of a cumulative repository of objective facts uninfluenced by the cultural or contextual and personal influences. Such an attempt utilized a closed system of intra personal or interpersonal deficits for the formulation of behavioral laws and principles. A close scrutiny of the vocabulary of psychological science shows that the social system and its power arrangements are neglected and individuals are overemphasized. "Psychological" became equivalent to the abstract concepts located inside the unique, separate, bounded, independent and context free individual and the assumed central processor inside it. Operating within the framework provided by the modernist spirit, the ideal human life is one arranged around the notion of individualism and the goal of possessing unlimited material resources. Similarly the idea of a good society was defined by the mechanisms of a market driven capitalist society.

The Analogical and Reductionist Strategies

Efforts to develop psychology along the lines of natural sciences have yielded either adoption of analogies or engaging with some form of reductionism. The analogies have often been taken from those, which are current in other disciplines. Interestingly to have an impact the idea needs time and by that time it may have been abandoned at its source. The fate of mechanistic models in physics are good examples, which are replaced with other models but continue to provide models for psychologists. The disciplines like chemistry, geology, mathematics, and economics have provided analogies to psychologists. In fact using analogies has been a major strategy to bring law and order into the world of psychology. The analogies or metaphors, once introduced, become part of the psychological reality. The fact that they were metaphors is often forgotten.

Another strategy to inject scienticism has been to reduce psychology to some other allegedly more developed science. It is held that the psychological phenomena are nothing but the expression at one level of mechanism and processes perfectly intelligible at another level. For instance Kohler considered mental processes as reflections of physical fields of force in the brain. Skinner viewed humans as nothing but an animal 'writ large'. Commenting on this situation Cohen (1970) cites the case of Tolman, who wrote *"I, in my future work intend to go ahead imagining how, if I were a rat, I would behave as a result of such and such a demand combined with such and such an appetite, and such and such a degree of differentiation and so on"*. Thus, to quote Cohen (1970) "he would imagine (thus employing a human capacity) how he would behave if he were a rat. On this precarious foundation he would explain the behaviour of the rat, and thence proceed to deduce an explanation for corresponding form of human behaviour. But if the exercise is based on imagining, why not make a short cut and study man directly "(p. 19). Cohen (1970) reminds us "the psychologist's task is not to cut his throat by writing himself off. It is to elucidate just those aspects of life, which have no place in the world picture of physical sciences, to begin in fact when the physicist leaves off. If psychology were allowed to make statements which could be translated into the language of neurophysiology, he would have nothing to say" (p. 30). Psychologists, however, are still fascinated with the use of these strategies and engage in constructing psychological realty in a natural scientific mode unmindful of its consequences.

Growth of Psychology and the Growing Discomfort

Today psychology is witnessing remarkable growth in quantitative terms. The number of journals, number of books published, number of practicing psychologists, organizations of psychologists and their professional bodies, and differentiation of the discipline is really mind-boggling. It is impossible to name all the specialities and sub-specialities of psychology. All this is really impressive for a new science. The growth of psychology in the developed nations of the world and developing countries, which have a history of colonization, show many parallels, if one goes by the themes, problems, methods, and findings. A majority of Indian psychologists, for instance, have dedicated their energy, time and money to replicate various phenomena in various combinations and permutations, the results of which, fortunately or unfortunately, are neither read nor used by our Indian colleagues in teaching or research. For our big brothers living in the developed countries this effort lacks any value and remains unregistered. There are a few exceptions, which have been able to receive some attention. On the whole the quality of research and publications is of mixed type (see for details Misra, Prakash, & Varma, 1999; Pandey, 2000). At times that gives a feeling that all is well. However, when we look for substance in terms of theoretical contribution, creativity or relevance the psychologists from developing countries often have to suffer from serious discomfort. The mechanical, Euro-American model of the so called mainstream psychology, has little to offer in terms of insights in the realm of theory and or practice which can be reassuring and inspiring. We suffer from lack of self-confidence, and academic self worth, and often engage in meaningless rote memorization of things having limited resemblance with our real world. The questions of authentic understanding or growth of knowledge are hardly addressed to, because as a true scientist we limit ourselves to the "context of justification", and the "context of discovery" remains ignored. The *status quo* is maintained and in the name of doing science the intellectual courage needed for questioning innovative thinking and creativity is given a back seat. The rituals of the religion called science continue.

The Challenge of Subjectivity and Focus on Instrumentality

The objectification of psychological processes by focusing on independent reality of objects, psychology's scientific paradigm denies subjectivity as a proper theme. The history of psychology shows that objectivity has been achieved at the cost of reflexivity. The move of humanistic psychology has also led to the creation of a technology of self. Also, the efforts have helped in the instrumentalization of psychology as a tool of facilitating State regulation. Development of many forms of applied psychology attests this. The subjectivity is stripped of its meaningful lived experience. This enterprise has been reactive in its stance and has been trying to evolve technologies to manage problems encountered without any effort to engage in the creative act of envisioning human condition and its psyche. The subject matter of psychology is distinctly human; it is not the mere lining of physiology. Our first step should therefore be to study what is characteristic of human beings, experiencing from the inside. Its subjectivity needs to be addressed and understood, and not explained away.

The Changing Scenario

In recent years many things have taken place and the situation in science, culture and psychology is changing. The notion of science is being considered as a human activity and culture is attended to in a serious manner. Alternate and multiple constructions are becoming visible. Heterogeneity in the discipline has started taking place and the possibility of a creative reconstructive is possible. Notable among these are the moves towards cross-cultural psychology, cultural psychology, and indigenous psychology. These moves, in different degrees have the capability to enrich the discourse on psychology. They favour a bottom up model of theorization and are transactional in nature. They view psychological world as an intentional world. The psychological phenomena are situated in meaningful contexts and their existence is intimately tied to the discursive practices. The interdependence of objects and persons is a cultural achievement. The scientific endeavours of the past are being debated and found constrained by congnitivism, individualism, artificiality. The decontextualized and abstract mode of psychology appears as irrelevant and unauthentic. Alternative paradigms include social representation, constructionist and discursive psychology, which emphasize the role of sharing, communicating, and negotiating which lead to a narrative construction of reality. The practice of using peer group as the place for reference and deriving legitimacy, coupled with internal consistency created by methodological practices have ruled the scientific endeavours. Now accountability, sharing and potential to deal with emergent properties are becoming relevant criteria. There is no transcendental realm of science. Psychology needs to be located in the day-to-day practices. It deals with values and deliberates on the issue of choice of values. Cultural criticism and contributing to cultural change are important tasks before psychologists.

Concluding Remarks

There is need to question psychology's self understanding as an expository of knowledge that explains the world yet is somehow detached from it. The alternative vision recognizes the complexity, uncertainty and contextual specificity. It recognizes a world where knowledge is itself socially constructed, fragmented, foundationless and validated by its usefulness rather than its scientific rigor (Gergen, 1999; Kvale 1992). Our nature of being is embedded within historical and cultural processes. However, the extra-discursive, material and embodied aspects of our being also need attention (Nightingale & Cromby, 1999). The debates on the place of culture in psychology (Berry, 1999; Misra & Gergen, 1993) and psychology in cultural context (European Psychologist, 2000; Gergen, Gulerce, Lock, & Misra, 1996) are drawing attention of psychologists. The *Asian Journal of Psychology* (December 2000) and *Journal of Cross-Cultural Psychology* (2001) have published special issues addressing the related concerns. The interface of *Psychology and Culture* is also being explored by a journal of the same name. The use of qualitative methods is also increasing. These methods are sensitive to culture and subjectivity and overcome the problem of power relations since they rely on participative, cooperative and constructive processes in the research process. It is hoped that these moves shall help in developing a human science of psychology in which cultural plurality will be addressed more authentically. The Indian thought system offers a unique opportunity to develop a comprehensive psychology rooted in one's relational existence, responsive to the multiple demands in a shared scheme of things. In a true sense, Indian thinking has the potential to attend to the issues of positive psychology (see Paranjpe, 1998). In 1965 in a paper in the *Journal of Humanistic Psychology* D. Sinha had invited the psychologists to work for the integration of modern psychology with Indian thought. In a recent assessment of the developments in Indian psychology J.B.P. Sinha (2000) observes that integrative indigenization is taking place. In future, more concerted efforts are expected to shape a vibrant, meaningful and relevant psychology through our sincere academic and social intervention.

References

Adair, J.G., Pandey, J., Begum, H.A., Puhan, B.N., & Vohra, N. (1995). Indigenization and development of the discipline: Perceptions and opinions of India and Bangladeshi psychologists. *Journal of Cross-cultural Psychology, 26*, 392-407.

Adair, J.G., Puhan, B.N., & Vohra, N. (1993). Indigenization of psychology: Empirical assessment of progress in Indian research. *International Journal of Psychology, 28*, 149-169.

Agrawal, A., & Tripathi, K.K., (1984). Influence of prolonged deprivation, age and culture on the development of future orientation. *European Journal of Social Psychology, 14*, 451-453.

Agrawal, A., Tripathi, K.K., & Shrivastava, M. (1983). Social roots and psychological implications of time perspective. *International Journal of Psychology, 18*, 367-380.

Akhilananda, S. (1948). *Hindu psychology*. London: George Allen & Unwin.

Allport, G.W. (1964). *The nature of prejudice*. Cambridge Mass: Wiley.

Anandlakshmy, S. (1974). How independent is the independent variable? In W.J. Dawson, & W. Lonner (Eds.) *Readings in cross-cultural psychology*. Hong Kong: University of Hong Kong Press.

Anandlakshmy, S., & Bajaj, M. (1981). Socialization among different cultures. In D. Sinha (Ed.), *Socialization of Indian Child* (pp. 31-38). New Delhi: Concept.

Anant, S.S., & Premshankar (1966). Inter caste differences in personality pattern, *Psychologia, 9*, 225-231.

Ardila, R. (1993). Latin American psychology and world psychology: Is integration possible? In U. Kim, & J.W. Berry (Ed.), *Indigenous psychologies: Research and experience in cultural context* (pp. 170-176). Newbury Park, CA: Sage.

Asian Journal of Social Psychology (2000). *Cross-cultural, cultural and indigenous psychologies*. December issue.

Atkinson, J.W., & Birch, D. (1978). *An introduction to motivation*. New York: D. Van Nostrand.

Bond, M.H. (1983). A proposal for cross-cultural studies of attribution. In Miles Hewstone (Ed.). *Studies in attribution*. Oxford: Basil Blackwell.

Bond, M. (1986). *Psychology of the Chinese People*. New York: Oxford University Press.

Brislin, R.W., Lonner, W.J., & Thorndike, R.M. (1973). *Cross-cultural Research Methods*. New York: Wiley.

Buss, A.R. (1975). The emerging field of the sociology of psychological knowledge. *American Psychologist, 30*, 988-1002.

Chakraborty, S.K. (1993). *Managerial transformation by values: A corporate pilgrimage*. New Delhi: Sage.

Cohen, John, (1970). *Homopsychologus*. London: George Allen & Unwin.

Cole, M., Gray, J., Glick, J.A., & Sharp, D.W. (1971). *The cultural context of learning and thinking*. New York: Basic Books.

Dalal, A.K. (1990). India. In G. Shouksmith & E.A. Shouksmith (Eds.) *Psychology in Asia and Pacific*. Bangkok: Unesco.

Dalal, A.K. (1996). A science in search of its identity: Twentieth century psychology in India. *Indian Psychological Abstracts and Review, 3*, 201-244.

Dalal, A.K., Singh, A.K., & Misra, G. (1988). Reconceptualizing achievement behaviour: a cognitive approach. In A.K. Dalal (Ed.) *Attribution Theory and Research*. New Delhi: Wiley.

Diaz-Guerrero, R. (1982). The psychology of historic socio-cultural premises. *Spanish Language Psychology, 2*, 282-310.

Doi, T. (1986). *The anatomy of self: The individual versus society*. Tokyo: Kodansha.

Dube, S.C. (1988). *Modernization and development: Search for alternative paradigm*. New Delhi: Vistar.

Dumont, L. (1970). *Homo hierarchicus: The caste system and its implication* (Trans. Maek S.) Chicago: Chicago University Press.

Durojaiya, M.O. (1993). Indigenous psychology in Africa: The search for meaning. In U. Kim , & J.W. Berry (Ed.), *Indigenous psychologies: Research and experience in cultural context* (pp. 211-220). Newbury Park, CA: Sage.

Endler, N.S., & Magnusson, D. (1976). Towards an interactional psychology of personality. *Psychological Bulletin, 83*, 956-74.

Enriquez, V.G. (1982). *Towards Filipino psychology: Essays and studies on language and culture*. Quezon City: Psychological Research & Training House.

Enriquez, V.G. (1992). *From colonial to liberation psychology: The Philippine experience*. Quezon City: University of the Philippines Press.

European Psychologist (2000). Quo vadis psychologia Europia, (*Perceptions assessments at the turn of the millennium*). Volume 5 number 2.

Fletcher, G.J. (1984). Psychology and common sense. *American Psychologist, 39*, 203-213.

Fox, D., & Prileltensky, I. (1997). *Critical psychology: An Introduction.* London: Sage.

Fromm, E. (1976). *To have or to be?* New York: Harper & Collins.

Gergen, K.J. (1973). Social psychology as history. *Journal of Personality and Social Psychology, 26*, 309-320.

Gergen, K.J. (1985). The social constructionist movement in modern psychology. *American Psychologist, 44*, 266-275.

Gergen, K.J. (1990). Therapeutic professions and the diffusion of deficit. *The Journal of Mind and Behaviour, 11*, 3-4.

Gergen, K.J. (1999). *An invitation to social construction.* London: Sage.

Gergen, K.J., Gulerce, A., Lock, A., & Misra, G. (1996). Psychological science in cultural context. *American Psychologist, 51*, 496-503.

Gupta, R.K. (1999). The truly familial work organization: Extending the organizational boundary to include employees' families in the Indian context. In H.S.R. Kao, D. Sinha, & B. Wilpert (Eds.), *Management and Cultural Values* (pp. 103-120). New Delhi: Sage.

Gupta, S.P. (1977). *Psychopathology in Indian medicine (Ayurveda) with special reference to its philosophical bases.* Aligarh: Ajaya Publishers.

Hall, C.S., & Lindzey, G. (1978). *Theories of Personality.* New Delhi: Willey Eastern Limited.

Harre, R., Clarke, D., & Decarlo, N. (1985). *Motive and Mechanisms.* London: Methuen.

Hassan, M.K. (1976). Self-image, social prejudice and child rearing practices, *Asian Journal of Psychology and Education, 12*, 30-37.

Hassan, M.K. (1978). A study of caste image, caste attitude and religiosity in high caste and scheduled caste college students. *Journal of Social Researches, 19*, 56-63.

Hassan, M.K. (1983). Parental influence on children's prejudice. *Social Change, 13*, 40-46.

Heider, F. (1958). *Psychology of interpersonal relations.* New York: Wiley.

Hendrick, C. (1977). Social psychology as an experimental science. In C. Hendrick (Ed.) *Perspectives on Social Psychology.* Hillsdale, N.J.: Lawrence Erlbaum.

Henley, N.M. (1986). Women as a social problem. In E. Seidman, & J. Rappaport (Eds.) *Redefining Social Problems.* New York: Plenum Press.

Hoshino, A., & Umemoto, T. (1987). Japanese psychology: Historical review trends. In Blowers, & A. Turtle (Eds.), *Psychology moving east: The status of western psychology in Asia and Oceania*, Boulder: West View Press, pp. 183-196.

Hurme, H. (1997). Psychological concepts, their producers and consumers. *Culture and Psychology*, 3, 115-136.

Jain, U. (1979). *Extreme response style as personality factor.* New Delhi: National.

Jain, U. (1990). Social perspectives on causal attribution. In G. Misra (Ed.) *Applied Social Psychology in India.* New Delhi: Sage.

Jain, U. (1994). Socio-cultural construction of emotion. *Psychology and Developing Societies, 6,* 151-168.

Jain, U., & Mal, S. (1984). Effects of prolonged deprivation on attribution of success and failure. *Journal of Social Psychology, 124,* 143-149.

Jain, U., & Misra, G. (1986). On the socio-cultural context of methodology. *Indian Journal of Community Guidance Service, 3,* 49-56.

Jones, J.M. (1990). Racism: A cultural analysis of the problem. In R.L. Jones (Ed.), *Black Psychology* (3rd Ed.), Berkeley: Cobb & Henry.

Kakar, S. (1978). *The inner world.* Delhi: Oxford University Press.

Kakar, S. (1995). *Culture and Psyche.* New Delhi: Oxford University Press.

Kakar, S. (1996). *The Indian psyche.* New Delhi: Oxford University Press.

Kao, H.S.R., & Sinha, D. (1997). *Asian perspectives on psychology.* New Delhi: Sage.

Kelley, H.H. (1967). Attribution theory in social psychology. In D. Levine (Ed.) *Nebraska symposium on motivation.* Lincoln: University of Nebraska Press.

Kim, U., & Berry, J.W. (1993). Introduction. In U. Kim, & J.W. Berry (Eds.), *Indigenous psychologies: Research and experience in cultural context* (pp. 1-29). Newbury Park, CA: Sage.

Kim, U., Park, Y., & Park, D. (2000). The challenge of cross-cultural psychology: The role of indigenous psychologies. *Journal of Cross-cultural Psychology, 31,* 61-75.

Kitayama, S., & Markus, H.R. (1999). Yin and Yang of the Japanese self: The cultural psychology of personality coherence. In D. Cervone, & Y. Shoda (Eds.), *The coherence of personality: Social-cognitive bases of consistency, variability, and organization* (pp. 242-302). New York: Guildford Press.

Koch, S. (1981). The nature and limits of psychological knowledge. *American Psychologist, 36,* 257-69.

Krishnan, B. (1974). Typological conceptions in ancient Indian thought.

In B. Krishnan (Ed.) *Studies in Psychology*. Mysore: University of Mysore.

Krishnan, L. (1997). Distributive justice in the Indian perspective. H.S.R. Kao & D. Sinha (Eds.), *Asian perspectives on psychology* (pp. 25-39). New Delhi: Sage.

Kuhn, T. (1970). *The structure of scientific revolutions*. Chicago: Chicago University Press.

Kukla, A. (1989). Nonempirical issues in psychology. *American Psychologist*. *44*, 785-794.

Kulkarni, T.R. (1984). A blueprint for developing an Indian School of Psychology. *Bombay Psychologist*, *56*, 18-34.

Kuppuswamy, B. (1975). *Social change in India*. Delhi: Vikas.

Kurtz, S.J. (1992). *All mothers are one: Hindu India and cultural shaping of psychoanalysis*. New York: Columbia University Press.

Kvale, S. (1992). *Psychology and postmodernism*. London: Sage.

Langer, E.J. (1983). *The psychology of control*. Beverly Hills California: Sage.

Leahey, T.H. (1991). *A history of modern psychology*, New Jersey: Prentice Hall.

Ma, H.K. (1972). The Chinese perspective on the moral judgment development. *International Journal of Psychology*, *23*, 201-227.

Mal, S., Jain, S., & Yadav, K.S. (1990). Effect of prolonged deprivation on learned helplessness. *Journal of Social Psychology*, *130*, 191-198.

Mamali, C. (1982). Democratization of the social research process. In P. Stringer (Ed.) *Confronting social issues: Application of social psychology* (Vol. 2) London: Academic Press.

Markus, H.R., & Kitayama, S. (1991). Culture and the self: Implications for cognition, emotion, and motivation. *Psychological Review*, *98*, 224-253.

Marriott, M. (1976). Hindu transactions: directions without dualism. In B. Kapferer (Ed.), *Transactions and meaning: Direction in the anthropology exchange and symbolic behaviour* (pp.109-142). Philadelphia: Institute for the Study of human Issue.

Marriott, M. (Ed.)(1990). *Indian through Hindu categories*. New Delhi: Sage.

Marsella, A.J. (1998). Toward a "Global-Community Psychology": Meeting the needs of a changing world. *American Psychologist*, *53*, 1282-1291.

Maruyama, M. (1974). Paradigms and communication. *Technological Forecasting and Social Change*, 6, 3-32.

Maxwell, N. (1984). *From knowledge to wisdom*, London: Basil Blackwell

Miller, J.G., & Bersoff, D.M. (1992). Culture and moral judgment: How are conflicts between justice and friendship resolved? *Journal of Personality Social Psychology, 62,* 541-554.

Misra, G., & Gergen, K.J. (1993b). Beyond scientific colonialism: A reply to Poortinga and Triandis. *International Journal of Psychology, 28,* 251-254.

Misra, G. (1982a). Motivational correlates of social disadvantage. In S. Dube, & P.S. Sachdeva (Eds.). *Mental health problems of the socially disadvantaged.* New Delhi: Tata McGraw-Hill.

Misra, G. (1982b). Experiential deprivation and development of representational competence. In R. Rath, H.S. Asthana, J.B.P. Sinha, & D. Sinha (Eds.). *Diversity and unity in cross cultural psychology* (pp. 70-77). Lisse, Netherland: Swets & Zeitlinger, B.N.

Misra, G. (1983). Deprivation and development. *Indian Educational Review, 18,* 12-32.

Misra, G. (1984a). Effects of disadvantage on attainment: The Indian Experience. *Bombay Psychologist, 5,* 102-114.

Misra, G. (1984b). Some recent trends in research on human motivation. *DIE Research Journal of Education and Psychology, 2,* 6-13.

Misra, G. (1985). Situational and personal determinants of attributions for achievement outcomes in India. *Indian Journal of Community Guidance Service, 2,* 1-18.

Misra, G. (1988). Social disadvantage and achievement attribution. In A.K. Dalal (Ed.) *Attribution Theory and Research* (pp. 81-97). New Delhi: Wiley Eastern.

Misra, G. (1990a). Psychology of deprivation. In G. Misra (Ed.) *Applied Social Psychology in India.* New Delhi: Sage.

Misra, G. (1991). Socio-cultural influences on moral behaviour. *Indian Journal of Social Work, 52,* 179-194.

Misra, G. (Ed.) (1990b). *Applied social psychology in India.* New Delhi: Sage.

Misra, G., & Agrawal, R. (1985). The meaning of achievement motivation: Implications for a cross-cultural theory of achievement motivation. In Reyes Lagunes I., & Y.H. Poortinga., (Eds.) *From a different perspective: Studies of behaviour across cultures.* Lisse: Swets and Zeitlinger.

Misra, G., & Gergen, K.J. (1993a). On the place of culture in psychological science. *International Journal of Psychology, 28,* 225-243.

Misra, G., & Giri, R. (1995). Is Indian self predominantly interdependent? *Journal of Indian Psychology, 13(1),* 16-29.

Misra, G., & Misra, S. (1986). Effect of socio-economic background on pupils' attributions. *Indian Journal of Current Psychological Research, 1,* 77-78.

Misra, G., & Shukla, A. (1986). Environmental deprivation and the development of spatial representation in children. *Psychologia, 29,* 115-122.

Misra, G., & Verma, S. (1995). Social problems, psychology and social policy. *Journal of Indian Academy of Applied Psychology, 20,* 97-105.

Misra, G., Prakash, Anand, & Varma, S. (1998). Psychology in India: Perceptions and perspectives. *Psychology and Developing Societies, 11,* 25-51.

Misra, V.N. (1984). *The image of man in Indian literature.* Personal Communication.

Mitra, S.K. (1972). *A survey of research in psychology.* Bombay: Popular.

Mohanty, A.K. (1988). Beyond the horizon of Indian psychology. The Yankee Doodler, In F.M. Sahoo (Ed.). *Psychology in Indian context* (pp. 1-8) Agra: National Psychological Corporation.

Moore, J.W. (1973). Social constraints on sociological knowledge: Academics and research concerning minorities. *Social Problems, 21,* 65-77.

Mukherjee, B.N. (1980). Psychological theory and research methods. In U. Pareek (Ed.), *A survey of research in psychology,* 1971-76, Part I, Bombay: Popular, pp. 1-135.

Murphy, G., & Murphy, L.B. (1968) (Eds.) *Asian psychology.* New York: Basic Books.

Naidu, R.K., Thapa, K., & Das, M.M. (1986). *On measuring detachment: An example of a scientific analog of an indigenous concept.* Unpublished manuscript. Department of Psychology, Allahabad University, Allahabad.

Naidu, R.N., Sethumadhavan, P., & Pandey, N. (1988). The state of psychology and national policies: introspection. Symposium paper Psychology, National Development and Social Policy Department of Psychology, Allahabad University, Allahabad.

Nakane, C. (1970). *Japanese Society.* Berkeley: University of California Press.

Nandy, A. (1974). The non-paradigmatic crisis in Indian psychology: Reflections on a recipient culture of science. *Indian Journal of Psychology, 49,* 1-20.

Nandy, A. (1983). *The intimate enemy: Loss and recovery of self under colonialism.* Delhi: Oxford University Press.

Nandy, A., & Kakar, S. (1980). Culture and personality. In Udai Pareek (Ed.) *A survey of research in psychology,* Bombay: Popular Prakashan, pp. 136-167.

Natraj, P. (1962). Stereotypes of four Hindu castes about each other. *Journal of Psychological Research, 6*, 132-141.

Neisser, U. (1978). Memory: What are the important questions? In M.M. Gruneberg, P.E. Morris & R.N. Sykes (Eds.) *Practical aspects of memory*. London: Academic Press.

Neki, J.S. (1973). Guru-Chela relationship: The possibility of a therapeutic paradigm. *American Journal of Orthopsychiatry, 43*, 755-766.

Nightingale, D.J., & Cromby, J. (1999). *Social constructionist psychology: A critical analysis of theory and practice*. Buckingham: Open University Press.

Nsamenang, A.B. (1992). *Human development in cultural context: A third world perspective*. Newbury Park, CA: Sage.

Nuttin, J. (1984). *Motivation, planning and action*. New Jersey: Erlbaum.

Ogedengbe, R.O. (1993). Some contributions of the traditional psychiatrists towards mental health in Nigeria. *Psychologia, 1*, 17-31.

Pandey, J. (1988). Psychology in India: Emerging trends. In J. Pandey (Ed.), *Psychology in India: The State of the Art (Vol. 3)*, New Delhi: Sage, pp. 339-355.

Pandey, J. (1988). *Psychology in India: The State of the Art (3 Volumes)*. New Delhi: Sage.

Pandey, J. (1990). *Perspectives on understanding social reality*. New Delhi: Concept.

Pandey, J. (2000). *Psychology in India revisited: Developments in the discipline*. New Delhi: Sage.

Pandey, N., & Naidu, R.K. (1992). Anasakti and health: A study of non-attachment. *Journal of Psychology and Developing Societies, 4*, 89-104.

Paranjpe, A.C. (1970). *Caste prejudice and the individual*. Bombay: Lalvani.

Paranjpe, A.C. (1984). *Theoretical Psychology: The meeting of East and West*. New York: Plenum Press.

Paranjpe, A.C. (1998). *Self and identity in modern psychology and Indian thought*. New York: Plenum Press.

Paranjpe, A.C., Ho, D.X.F., & Ribber, R.W. (Eds.) (1988). *Asian contributions to psychology*. New York: Praeger.

Pareek, U. (1980). *A survey of research in psychology*, 1971-76. Bombay: Popular.

Pepitone, A. (1989). Toward a cultural social psychology. *Psychology and Developing Societies, 1*, 5-19.

Popper, K.R. (1967). The sociology of knowledge. In K.R. Popper (Ed.) *Open society and its enemies*. New York: Harper & Row.

Popper, K.R., & Eccles, J.C. (1977). *The self and its brain: An argument for interactionism*. New York: Springer Verlag.

Prabhu, P.H. (1963). *Hindu social organization: A study in social psychological and ideological foundation*. Bombay: Asia.

Prakash, A. (1994). Organizational functioning and values in the Indian context. In H. Kao, D. Sinha, & N. Sek Hong (Eds.), *Effective organizations and social values* (pp. 193-301). Delhi: Sage.

Prasad, M.B. (1973). Development of caste awareness in children: Theoretical model. *Journal of Socioeconomic Studies, 1*. 115-129.

Prigogine, I., & Stengers, I. (1984). *Order out of chaos*. London: Flamingo.

Prillettensky, I. (1989). Psychology and the status quo. *American Psychologist, 44*, 795-802.

Priolieu, L., Murdock, M., & Brody, N. (1983). An analysis of psychotherapy versus place of studies. *The Behavioral and Brain Sciences, 6*, 275-310.

Rai, I. (1990). Parents of prejudiced and unprejudiced Hindu school students: Their attitudes and personality. Unpublished doctoral dissertation, Ranchi University, Rachi.

Ramanujan, A.K. (1990). Is there an Indian way of thinking? An informal essay. In McKim Marriott (Ed.) *India through Hindu categories*. New Delhi: Sage.

Ramchandra Rao, S.K. (1962). *Development of psychological thought in India*. Mysore: Kavyalaya Publishers.

Rao, K.R. (1986). What is Indian psychology? *Journal of Indian Psychology, 7*, 37-57.

Rao, K.R. (1988). Psychology of transcendence: A study in early Buddhistic psychology. In A.C. Paranjpe, D.X.F. Ho, & R.W. Ribber (Eds.), *Asian contributions to psychology*. New York: Praeger.

Rath, R., & Das, J.P. (1957). Study in stereotype of college freshmen and service holders (Orissa) towards Indian groups. *Indian Journal of Psychology, 33*, 239-252.

Rath, R., & Sirkar, N.C. (1960). The mental pictures of six Hindu caste groups as reflected in verbal stereotypes. *Journal of Social Psychology, 51*, 277-293.

Rath, R., Das, A.S., & Das, U.N. (1979). *Cognitive abilities and school achievement of the socially disadvantaged children in primary schools*. Bombay: Allied.

Risley, H.H. (1969). *The peoples in India*. Delhi: Oriental Books.

Rolland, A. *(1988). In search of self in India and Japan*. New Jersey: Princeton University Press.

Safaya, R. (1976). *Indian psychology*. New Delhi: Munshi Ram Manoharlal.

Sampson, E.E. (1978). Scientific paradigms and social values: Wanted a scientific revolution. *Journal of Personality and Social Psychology, 36*, 1132-1143.

Sampson, E.E. (1978). Scientific paradigms and social values: Wanted a scientific revolution. *Journal of Personality and Social Psychology, 36*, 1132-1143.

Sampson, E.E. (1989). The challenge of social change for psychology: Globalization and psychology's theory of the person. *American Psychologist, 44*, 914-921.

Saraswathi, T.S. (Ed.). (1999). *Culture, socialization and human development: Theory research and applications in India*. New Delhi: Sage.

Sarkar, S.N., & Hassan, M.K. (1973). Economic conservatism as related to religion, caste, political affiliation and authoritarianism. *Indian Journal of Psychology, 48*, 64-70.

Seidman, E. (1978). Justice, values and social science: Undermined premises. In R.J. Semon (Eds.). *Research in law and sociology (Vol. 1)*. Greenwich, CT: JAI Press.

Seidman, E., & Rappaport, J. (1986). *Redefining social problems*. New York: Plenum Press.

Seymour, S. (1988). Expressions of responsibility among Indian children: Some precursors of adult status and sex role. *Ethos, 16*, 355-370.

Sharma, A., & Anandlakshmy, S. (1981). Prejudice in the making: understanding the role of socialization. In D. Sinha (Ed.). *Socialization of the Indian child* (pp. 106-116). Delhi: Concept.

Sharma, K.L. (1988). *Understanding philosophical concepts by the laymen*. Unpublished manuscript, Rajasthan University, Jaipur.

Sharma, R., & Tripathi, R.C. (1988). Teachers expectations and attributions. The self fulfilling prophecy cycle. In A.K. Dalal (Ed.). *Attributions theory and research* (pp.33-59), New Delhi: Wiley Eastern.

Sharma, S. (1981). Key concepts of social psychology in India. *Psychologia, 24*, 105-14.

Shweder, R.A. (1990). Cultural psychology—what is it? In W. Stigler, R.A. Shweder, & G. Herdt (Eds.), *Cultural psychology: Essays on comparative development* (pp.1-43). Cambridge: Cambridge University Press.

Shweder, R.A., Mahapatra, M., & Miller, J.C. (1987). Culture and moral development. In J. Kagon & S. Lamp (Eds.), *Emergence of moral concept in early childhood*. Chicago: Chicago University Press.

Singh, A.K. (1978). *Deprivation: A review of studies.* Unpublished manuscript. Psychology Department, Ranchi University, Ranchi.

Singh, A.K. (1980). *Prejudice in India: Development and correlates.* ICSSR Project Report, New Delhi.

Singh, A.K. (1988). Intergroup relations and social tension. In J. Pandey (Eds.). *Psychology in India. The state-of-the-art.* Vol. 2, (pp.159-224), New Delhi: Sage.

Singh, R., & Nathawat, S.S. (1989). Effect of prolonged deprivation on attributional style. *Journal of Social Psychology, 129,* 443-450.

Singh, S., & Misra, G. (1985). Effect of competence manipulation on cognitive performance and persistence in relation to socio-economic deprivation. *Psychological Studies, 30,* 12-26.

Singh, S., & Misra, G. (1988). Attribution for performance as a function of deprivation and ethnicity under competence and helplessness conditions. *DEI Research Journal of Education and Psychology, 6,* 1-7.

Singh, S.D., Singh, K., & Singh, Y. (1960). Development of caste consciousness in Indian children from 4 to 10 years. *Agra University Research Journal, 9,* 109-125.

Singh, Y. (1982). Relevance of social science in India. In P.C. Joshi (Eds.). *Relevance in social science research.* New Delhi: Vikas.

Singh, Y. (1989). *Social stratification and change in India.* New Delhi: Munshi Ram Manohar Lal.

Sinha, A.K.P. (1971). Relationship between ethnic stereotypes and social distance. *Psychological Report, 28,* 216.

Sinha, D. & Rao, H.S.R. (Eds.) (1988). *Social values and development: Asian perspectives.* New Delhi: Sage.

Sinha, D. (1965). Integration of modern psychology with Indian thought. *Journal of Humanistic Psychology, 5,* 6-17.

Sinha, D. (1969). *Indian villages in transition.* New Delhi: Allied Publishers.

Sinha, D. (1981). Non-western perspectives in psychology: Why what and whether? *Journal of Indian Psychology, 3,* 1-9.

Sinha, D. (1983). Human assessment in the Indian context. In S.H. Irvine and J.W. Berry (Eds.) *Human assessment and cultural factors.* New York: Plenum Press.

Sinha, D. (1984). Reflections on measurement of human motivation in India. *Psychological Studies, 29,* 197-205.

Sinha, D. (1985). Application of psychology in developing countries. *Chinese Journal of Applied Psychology, 1,* 9-15.

Sinha, D. (1986). *Psychology in a third world country: The Indian experience.* New Delhi: Sage.

Sinha, D. (1990). Concept of psychosocial well-being: Western and Indian perspectives. *NIMHANS Journal, 8,* 1-11

Sinha, D. (1994). Indigenous psychology: Need and potentiality. *Journal of Indian Psychology, 12,* 1-7.

Sinha, D. (1997), Indegenizing psychology. In J.W. Berry, Y.H. Poortinga & J. Pandey (Eds.), *Handbook of cross-cultural psychology: Theory and method, Vol. 1* (pp.129-169), Boston: Allyn and Bacon.

Sinha, D., & Misra, G. (1982). Deprivation and its motivational and personality correlates. In D. Sinha, R.C. Tripathi, & G. Misra (Eds.), *Deprivation: Its social roots and psychological consequences.* New Delhi: Concept.

Sinha, D., & Naidu, R.K. (1994). Multilayered hierarchical structure of self and not self: The Indian perspective. In A.M. Berry, F.J.R. Van de Vijver, P. Boski & P. Schmize (Eds.). *Journey into Cross-cultural psychology.* (pp.41-49). Lisse: Swets & Zeitlinger.

Sinha, G.S., & Sinha, R.C. (1967). Explorations in caste stereotypes. *Social Forces, 46,* 42-47.

Sinha, J.B.P. (1980). *The Nurturant-task leader: A model of the effective executive.* New Delhi: Concept.

Sinha, J.B.P. (1982). Concept of social energy. Paper presented at the Seminar on *Issues in Research on Human Motivation,* Department of Psychology, Allahabad University, Allahabad.

Sinha, J.B.P. (1984). Toward partnership for relevant research to the third world. *International Journal of Psychology, 19,* 169-178.

Sinha, J.B.P. (1988). *Developing psychology as a policy science: Prospects and problems.* Symposium paper: Psychology, National Development and Social Policy, Department of Psychology, Allahabad University, Allahabad.

Sinha, J.B.P. (1997). In search of my Brahman. In M.H. Bond (Ed.), *Working at the interface of culture: Eighteen lives in social science* (pp. 77-84). London: Routledge.

Sinha, J.B.P. (2000). Indigenization of psychology in India. *Psychological Studies, 45,* 3-13.

Sinha, J.B.P., & Pandey, J. (1970). Strategies of high n-Ach persons. *Psychologia, 13,* 210-216.

Sinha, J.B.P., & Prakash, A. (1993). *ICSSR's contributions to the growth of psychology in India.* New Delhi: ICSSR.

Sinha, S. (1992). *Caste system: Myths, reality and challenge,* New Delhi: Intellectual.

Smith, J.A., Harre, R., & Langerhove, L.V. (1985). *Rethinking psychology.* London: Sage.

Steenbergen, B.V. (1990). Potential influence of the holistic paradigm on the social science. *Future, 22,* 1971-1983.

Stringer, P. (Ed.) (1982). *Confronting social issues: Application of social psychology*. London: Academic Press.

Tajfel, H. (1974). Social identity and inter group behaviour. *Social Science Inference, 13,* 65-69.

Tajfel, H. (1978). *Differentiation between social groups*. London: Academic Press.

Tiwari, R.K. & Misra, G. (1980). Some observations on the development of caste awareness in children. *Indian Journal of Social Work, 39,* 4439.

Traindis, H.C. (1993). Comments on Misra and Gergen. *International Journal of Psychology,* 28.

Triandis, H.C. (1980). Introduction. In H.C. Triandis, & W.W. Lambert (Eds.). *Handbook as cross-cultural psychology (Vol. 1)*. Boston: Allya and Bacon.

Tripathi, L.B. (1982). Some methodological problems of deprivation studies. In D. Sinha, R.C. Tripathi, & G. Misra (Eds.) *Deprivation: Its social roots and psychological consequences* (pp. 49-71). New Delhi: Concept.

Tripathi, R.C. (1982). Concept of deprivation. In D. Sinha, R.C. Tripathi, & G. Misra (Eds.), *Deprivation: Its social roots and psychological consequences* (pp.17-23). New Delhi: Concept.

Tripathi, R.C. (1988). Aligning development to values in India. In D. Sinha & H.S.R. Rao (Eds.). *Social values and development: Asian perspectives* (pp. 315-333). New Delhi: Sage.

Verma, S. (1995). The social constructionist framework: An alternative paradigm for psychology. *Indian Journal of Social Science, 18.*117-129.

Vyas, S.K. (1973). The origin of prejudice in children. In T.E. Shanmugan (Ed.), *Research in personality and social problems.* Madras: University of Madras.

Wang, J.S. (1980). On the modernization of psychology in China: Fundamental theoretical viewpoint and methodology. *Acta Psychological Sinica,* 12, 30-36 (in Chinese). English translation: FBIS-REPORT Joint Publications Research Service, No. 76133, 103-114.

Weiner, B. (1986). *An attributional theory of motivation and emotion*. New York: Springe Verlog.

Wilpert, B. (1998). *Merging of past, present and future.* Presidential address to International Association of Applied Psychology, San Francisco.

Yamaguchi, S. (1994). Collectivism among the Japanese: A perspective from the self. In U. Kim, H.C. Triandis, C. Kagitcibasi, S.C. Choi & G. Yoon (Eds.), *Individualism and collectivism: Theory, method, and applications* (pp. 175-188). London: Sage.

Yang, K.S. (1982). Sinicization of psychological research in Chinese society: Directions and issues. In K.S. Yang & C.I. Wen (Eds.), *The sinicization of social and behavioral science research in China.* Taipei, Taiwan: Institute of Ethology, Academia Sincia (In Chinese), pp. 153-187.

Zafar, S., & Alam, K. (1985). Effect of different facets of prolonged deprivation on retention. *Psychological Studies, 31*, 26-28.

Zilbergeld, B. (1983). *The shrinking of America: Myths of psychological change,* Boston: Little Brown.

Index